AERIAL SURVEYING BY RAPID METHODS

AERIAL SURVEYING BY RAPID METHODS

by

BENNETT MELVILL JONES, A.F.C., M.A.

FRANCIS MOND PROFESSOR OF AERONAUTICAL ENGINEERING,
AND FELLOW OF EMMANUEL COLLEGE, IN THE
UNIVERSITY OF CAMBRIDGE

and

MAJOR J. C. GRIFFITHS, B.Eng., Hon. B.Sc.

CAMBRIDGE
AT THE UNIVERSITY PRESS
1925

CAMBRIDGE
UNIVERSITY PRESS

University Printing House, Cambridge CB2 8BS, United Kingdom

Cambridge University Press is part of the University of Cambridge.

It furthers the University's mission by disseminating knowledge in the pursuit of
education, learning and research at the highest international levels of excellence.

www.cambridge.org
Information on this title: www.cambridge.org/9781107511514

© Cambridge University Press 1925

First published 1925
First paperback edition 2015

A catalogue record for this publication is available from the British Library

ISBN 978-1-107-51151-4 Paperback

Additional resources for this publication at www.cambridge.org/9781107511514

AUTHORS' FOREWORD

THE way in which the series of experiments described and analysed in this book originated is explained in the preface by Captain H. Hamshaw Thomas, M.B.E.; it remains to explain the somewhat peculiar circumstances under which the experiments were made and the book written.

The experimental investigations on this subject, which began in Cambridge in the autumn of 1920, were the result of co-operation between many departments and individuals. Major J. C. Griffiths, who was at the time a research student in the University, carried out the bulk of the experiments, supported, meanwhile, by a financial grant from the Department of Scientific and Industrial Research. The Royal Air Force provided the aeroplanes and pilots together with cameras and other instruments, whilst the Royal Aircraft Establishment lent certain experimental apparatus which were not at the time standard in the Air Force. The direction of the experiments and the responsibility for the utilisation of the apparatus and facilities provided by these departments lay in my hands, but the original inspiration concerning the need for the experiments and the type of information required came from Captain Thomas, as the result of his experience in Palestine during the campaign of 1917–8. We have been in close touch with Captain Thomas throughout the course of the experiments and his advice and criticism have been invaluable.

During the whole of the work valuable assistance has been received, both from private individuals and from members of the staffs of the services and departments mentioned. In particular the Air Survey Committee, a joint committee of the three fighting services, has co-operated wholeheartedly. By the kindness of the Chairman, Colonel H. StJ. L. Winterbotham, C.M.G., D.S.O., I was invited to attend the meetings of this Committee and was kept informed of its activities, so that, together, it was possible to arrange a combined scheme of work which would cover, without undue repetition, the whole of the ground over which it appeared that experiments were required. The assistance rendered by the members of this Committee, both collectively and individually, has been very great, and it was mainly through their support that we were able to borrow the costly and delicate apparatus that we required.

The investigations could not have taken place at all without the material assistance rendered by the Royal Air Force. The Chief of the

Air Staff, Sir Hugh M. Trenchard, G.C.B., D.S.O., provided a number of aeroplanes for our use, detailing pilots to fly them and mechanics to keep them in working order. The small experimental flight so formed operated at the service aerodrome of Duxford, near Cambridge; it was placed absolutely at our disposal "except in so far as the exigencies of the service permitted." For administrative and disciplinary purposes this flight was under the control of the Officer Commanding the station, and our sincere thanks are due to Group Captain W. R. Freeman, D.S.O., M.C., and to Wing Commander Sidney Smith, D.S.O., A.F.C., for the sympathetic manner in which they interpreted this arrangement whilst they were in command of this station. Experimental work in the air, in the uncertain weather that prevails in this country, is at all times difficult to carry through successfully. The necessity of seizing every spell of fine weather, regardless of times and seasons, makes such work particularly difficult to fit in with the ordered routine of service life; it is not too much to say that no appreciable progress could have been made had not the Commanding Officers taken a personal interest in the work and done all in their power to push it forward.

The piloting throughout the experiments was in the hands of three officers: Flight Lieutenant F. H. Coleman and Flying Officer C. E. H. Allan, D.F.C., working together for the greater part of the time, and Flight Lieutenant D. L. Blackford working alone in the later stages. The work of these officers will appear in the text, but what cannot be indicated there is the enthusiasm with which they carried out their instructions, which often involved long and very arduous flights at times when they would normally be off duty. I wish also to take this opportunity of thanking Flying Officer D. F. Drudge, who was for some time in charge of the photographic section at Duxford, for the enthusiastic way in which he co-operated and the interest that he took in the work, and I feel that special mention must be made of two air mechanics, Smallman and Bollen, whose enthusiastic work as engine man and rigger respectively contributed materially to the success of the experiments.

The more elaborate and expensive instruments necessary for the experiments were received on loan from the Royal Air Force and the Royal Aircraft Establishment, and in connection with these loans I have to thank particularly Squadron Leader F. C. V. Laws, C.B., O.B.E., in charge of the Photographic Department of the R.A.F., Mr W. Sydney Smith, C.B., O.B.E., Superintendent of the R.A.E., Mr G. L. Smith the head of the Instrument Section of the R.A.E. and Major H. E. Wimperis head of the Navigation Department of the Air Ministry.

Finally help and advice has been received from many private individuals, amongst whom I would like to tender special thanks to Dr G. T. Bennett, F.R.S., for assistance on the geometrical side, Mr F. Debenham, O.B.E., for help on the surveying side, and my wife for help in connection with the reduction of data and the typing of the manuscript.

In the spring of 1923 the experiments had reached a stage when we judged that they could conveniently be collected and published in the form of a short book, and an agreement was concluded with the Cambridge University Press for the publication of a book, under the joint authorship of Major Griffiths and myself. Certain experimental work, connected with the scheme of Navigational Control combined with oblique photographs, remained to be completed, but we judged that this would not take very long, and the experiments already completed indicated clearly the nature of the results to be expected. At this juncture Major Griffiths received the offer of a very responsible post with a firm of Aircraft Constructors. We considered that this offer was too good to be missed, particularly as the firm had agreed to extend to him facilities for the completion of his experiments, and to release him for part time in the winter to allow the results to be analysed, and to enable him to take his part in the writing of the book. In the meantime, during the summer of 1923, I was to take his place as observer at Duxford and carry out what experiments I could. We realised, however, that I should be able to devote a part of my time only to this work, and that therefore there was little hope of my getting much done: we relied principally upon Griffiths being able to finish off the work, either in his new situation, or whilst on leave.

The book was to have gone to press about Christmas 1923, but in October of that year Griffiths was accidentally killed whilst flying, and thus all our plans were dislocated.

After Griffiths' death the final analysis of much of the experimental data and the writing of the book itself fell entirely on my shoulders, and as I could only give a portion of my time to the work, its completion was much delayed, so that publication is likely to occur nearly a year later than was originally intended. The situation so created placed me in considerable difficulty. I was anxious that Griffiths' name should remain as joint author, because many of the ideas and most of the experiments were his, but it was clear that I should have to write it myself and even draw some of the final conclusions without Griffiths' assistance or approval. I therefore felt, and still feel, some diffidence in expressing, under his name, conclusions with which conceivably he might not have agreed; I could not have adopted this course at all had not the main conclusions and the general form of the book been agreed upon between us before his accident.

I think I am right in saying that the first five chapters contain only matter which had been discussed between us so thoroughly that there is no danger of my having inserted anything in relation to which we were not in complete agreement; but, in regard to the last four chapters, this cannot be said with the same degree of certainty. The ideas involved in chapter VI, dealing with Navigational Control and the use of oblique photographs, had been extensively discussed between us, but much of the data had either not been obtained, or had not been analysed, at the time of our last discussion. Although, therefore, the general outlines of this chapter had been agreed between us, I must take the responsibility for the detailed conclusions reached.

Chapter VII, on detailed routine, was to have been written, in rather longer form, by Griffiths himself, but he had apparently not been able to start upon it up to the time of his accident, and I have had to do my best with it from my own experience and memory of our previous discussions. In general it may be said to be in conformity with ideas that we held in common.

For some reason we had omitted to discuss the necessity for the inclusion of a chapter on the training of personnel. Some reference to this question was clearly necessary to complete the work, and I therefore wrote chapter VIII, as nearly as I could in conformity with our common views; here again much of the data upon which the detailed conclusions are founded had not been analysed at the time of his death, although the general trend of the results was sufficiently evident.

For the final chapter, which summarises the work and draws conclusions from it, I must take entire responsibility, although I have no doubt that Griffiths would have been in agreement with the general outlines.

I feel I cannot close this foreword without some reference to Griffiths himself. A man of immense courage and enthusiasm, he was exactly suited to the work upon which he was engaged for the last few years of his life. He had that rare temperament which fits a man for active and dangerous work out of doors, and yet does not prevent him from excelling in the laboratory and office. It is certain that the aeronautical world in general, and aerial surveying in particular, has suffered a great loss by his death. His accident re-opens the old controversy concerning the extent to which scientists of exceptional promise should be allowed and encouraged to take part in dangerous occupations. The problem is really one of degree, and all that can be said about the case in question is that there must, at all times, be some individuals filling the difficult rôle of intermediary between the theoretical and laboratory world, and the world of practical aeronautics, and that rôle was one for which Griffiths was particularly fitted. This he realised and faced the risk joyfully, as an integral part of his life. Of all men who have aspired to fill this rôle I have known none who stood a better chance of coming safely through the risks involved, but the nature of accidents is that they cannot, in general, be foreseen and may happen to anyone, no matter how skilful he may be.

Young though he was, Griffiths had already contributed greatly to our knowledge of a subject in which experiment is by no means an easy matter; by his energy and application to work he had achieved results which will, I believe, be found to have a permanent value. To the natural sympathy which we all feel, in such circumstances, for those who have lost the society of a delightful and stimulating personality, is added our deep regret for the loss of that much greater volume of pioneer work which he would undoubtedly have carried through had his life been longer.

B. M. J.

PREFACE

By CAPTAIN H. HAMSHAW THOMAS

ONE of the characters of the modern world is its dependence on maps. Scarcely a day passes without the appearance in the press of a map illustrating the site of some current event, while the millions of travellers, hurrying to and fro on the earth in trains, boats or motor vehicles, are for the most part intelligent map-readers. At the same time, in every part of the world, schemes are on foot for the more effective utilisation of the earth's surface, and the first step towards the carrying out of such schemes is the preparation of a map or plan.

Those who live in a small country like England, which has been accurately mapped in great detail, can scarcely realise the difficulties and drawbacks of life in a country whose maps are of an imperfect description. Travel becomes a matter of difficulty and uncertainty, the development of agricultural land and the utilisation of forests are retarded, the exploitation of minerals may have to be long postponed. In some places many small surveys are made from time to time, but they are generally the outcome of private effort and seldom benefit the community as a whole.

As time goes on the demand for a complete topographical survey, such as exists for the countries of Western Europe, will probably become irresistible in other settled parts of the world, and with the march of civilisation it cannot be ultimately avoided. But in the vast territories of Africa, Australia and America a topographical survey of the ground in any detail is a colossal undertaking and would involve an enormous expenditure of time and money. Moreover in some places forests, mountains or other difficulties render the task of surveying ground-detail almost impossible. The expense of keeping survey parties in the field long enough to secure the desired results would be so great that the production of the desired maps may be indefinitely postponed. But at the same time we find that trigonometrical surveys are now being made in many places. The geographical position of many isolated points is being determined and a foundation is being laid for the topographical surveys.

It would seem that we need some new survey method which has the power of depicting very rapidly the surface features of the country in considerable detail, which can be operated over a wide tract of country, far from bases or settlements, and which is unhampered by forests, mountains or broken country.

During the war topographical maps were required of many places which had not been surveyed in any great detail, and on the Palestine Front the writer was intimately concerned with the construction of maps of inaccessible country by the aid of aeroplane photography. A wide area—over 2000 square miles—had to be covered, the country was very broken, and the greatest possible detail was required. A certain number of points had been fixed by previous surveyors but these were generally many miles apart. By the joint efforts of officers belonging to the Royal Air Force and to the Royal Engineers with the most valuable help of some members of the Survey of Egypt, a system was evolved by which a series of maps was produced which were found to be of considerable utility and of fair accuracy. The conditions of the work, the needs of the situation and the methods employed, differed considerably from those which were evolved independently on the Western Front.

This system, which had been elaborated empirically to meet the needs of the situation, seemed to possess some of those features indicated above as desirable in a new method of survey. It appeared to have the advantages of speed, range of action and wealth of detail, but it was not definitely known to what extent such a survey would be sufficiently accurate and economical for civil use.

An accurate map should show correctly the exact geographical position as well as the correct shape of all ground features, but in most topographical maps made by the ordinary methods there is some departure from the theoretical accuracy. When they are compared with corrected air-photographs or with maps made by other photographic processes, it is found that the surveyor on the ground has often to make generalisations and to be content with approximate shapes and positions. To the map user, accuracy of shape and correct relative position may at times be more important than correctness of geographical position with generalisation of shape.

While the Palestine methods of topographical survey by aerial photography seemed to have possibilities of great utility, it was based on the assumption that aeroplanes could be flown for a certain distance at a uniform height and with their wings level. It was also inapplicable to mountainous country, and was more especially designed for territory in which a complete knowledge of all the surface features in detail was desired.

Before attempting to apply this method for civil uses, it was necessary to investigate whether the underlying assumptions were correct, to determine the limits of its accuracy and to consider whether it could be used as the foundation of methods applicable to mountainous or desert countries. At this juncture the late Major Griffiths arrived at Cambridge, having taken his degree in Engineering at Liverpool University and possessing

Preface xiii

considerable flying experience. He was full of enthusiasm and was anxious to investigate the possibilities of aerial photographic survey and the chances of using it for the exploration and development of distant unmapped lands overseas. To his work this volume owes its origin. Undeterred by laborious and often fruitless labour, anxious to establish firm foundations, patient under the difficulties of working with inadequate materials and machines designed for totally different purposes, he carried out with Prof. Melvill Jones a series of researches which are of fundamental and lasting importance. It is a matter of great regret that he did not live to complete his work and to see his results utilised. The fatal accident at Coventry on October 20, 1923 has deprived the world of one of its most talented investigators in the sphere of aeronautics.

The researches, described in the following pages, come at a very opportune moment. The last five years have seen the formation of aeroplane units in various parts of the world which are in need of extended topographical surveys. In the British Empire the Dominion of Canada, the Commonwealth of Australia and the Union of South Africa each has its Air Force, and all of these organisations have shown an interest in aerial photographic surveying. In Canada, where pioneer work had been done in photo surveying on the ground, several surveys have already been made from the air. The Canadian Air Board reported in 1923 that experimental work was being carried out and that "Important progress has already been made and a successful conclusion may mean much to those branches of the Government service engaged in the mapping of Canada."*

In Canada it has been found that aerial photographic survey, while of great value for mapping the boundaries of forests and the waterways running through them, is also of great utility for the location of the more valuable timber trees. Thus in a forest survey two distinct services are performed at the same time. This illustrates the fact that such a survey depicts not only the surface features of the ground but also the different types of vegetation growing upon it, or the beds of rock which may be exposed on the surface. The vegetation has often a most important economic significance in the development and colonisation of a new country, as it gives indications of the presence of springs and of water supply, apart from its own intrinsic value in wooded countries. Likewise the location of rock outcrops may be of great value to the geological surveyor and in the investigation of mineral resources.

The reports of the Canadian Air Board show how aerial photo surveying may have still other useful practical results, in addition to map production, and emphasise the potentialities of the method. A series of

---

* *Report of the Air Board for* 1922, Ottawa, 1923, p. 50.

air photos has a permanent value, and after it has been used for purely topographical purposes it may be kept for reference and used from time to time as required, in connection with land settlement or irrigation schemes, in road and rail construction, for vegetation or geological surveys and so on. While there seems to be very little scope for aerial survey in England, it may be a means of facilitating the progress of civilisation in many lands overseas.

It has always seemed to me that the future of aerial survey lay in its ability to do work which could not be readily accomplished by the usual methods on the ground, work in which the nature of the terrain, the wide area to be covered, or its inaccessibility were limiting factors. I therefore suggested to Prof. Melvill Jones and Major Griffiths that it would be useful to investigate the relations existing between aerial methods and ground control. It is always necessary to have some points in the area whose position has been fixed by trigonometrical survey, but the profitable employment of aerial photography will be greatly limited if a large number of such points are required. If, for example, it is necessary to have three fixed points on every photograph, we shall be able to do little in some forest lands. I also pointed out that where a wide area has to be dealt with, it is necessary to reduce as far as possible the labour of compiling a map from the photographs; therefore it might be well to avoid methods which necessitated the accurate measurement of a large number of distances and angles on the prints or negatives.

In acting on these suggestions the investigators have travelled along lines which are very different from those of most of the other workers on the subject, and in so doing they have contributed results of great importance. They have not, of course, arrived at any final and complete scheme, but they have laid foundations, and have gone far towards producing methods which have great possibilities. They dealt with flat country, but some of their work may be developed for use in hilly country. They have not investigated the application of stereoscopic methods, but their results on the investigation of the path of a machine through the air indicate that success may be expected in the survey of mountains and the measurement of heights by the use of oblique stereoscopic photography.

It is certain that the investigations described in this volume constitute an important step in the evolution of the methods of aerial survey, and will eventually result in great additions to our knowledge of the surface of the earth.

CONTENTS

PAGE

AUTHORS' FOREWORD V

PREFACE xi
By *Captain H. Hamshaw Thomas*

CHAP.
   I. INTRODUCTION 1

  II. THE PROBLEM OF PHOTOGRAPHING THE WHOLE GROUND . 7

 III. THE PROBLEM OF MAKING A MAP FROM THE PHOTOGRAPHS . 17

 IV. PRELIMINARY EXPERIMENTS 31

  V. MAPPING BY MOSAICS OF VERTICAL PHOTOGRAPHS . . 47

 VI. MAPPING BY NAVIGATION AND OBLIQUE PHOTOGRAPHS . 67

VII. EQUIPMENT AND DETAILED PROCEDURE 103

VIII. TRAINING 124

 IX. CONCLUSIONS 131

APPENDICES
   I. The Methods used in the Study of Random Errors 140

  II. Pilots Reports of the flights discussed in Chapter IV 143

 III. Part I. On the Influence of a Steady Rate of Change of Wind upon the
      Accuracy of Navigational Control 145
     Part II. On the Elastic Adjustment of Errors in a Navigationally Controlled
      Survey 146

 IV. On the Construction of Graticules for the Measurement of Relative Azimuth
     Angles and Depressions below the Horizon 148

  V. On the Probability of Error in the Estimation of Height by the Methods of
     Chapter VI 152

 VI. On the Correction of the Air-speed Indicator for Variations in Air Density . 155

VII. On the Wind Gauge Bearing Plate and its use in Aerial Surveying . . 157

LIST OF ILLUSTRATIONS

FIGURES IN TEXT

FIG.   PAGE

1.   Curves of approach to an objective 12

2.   Specimen form for Observer 119

3.   Specimen form for Pilot 120

PLATES AS INSETS

The D.H. 9 A . . . . FACE 34

Showing photographic strips with fictitious curvature. Showing process of compilation of mosaic without independent control . . . . . . ,, 54

The mosaic on the compilation table showing process of compiling mosaic with
independent control. Showing, above an "indication strip," below the
effect of omitting to rotate the camera to allow for "drift" . . . . ,, 55

Oblique photograph taken from 10,000 feet. Oblique photograph with graticule . ,, 90

Showing pilot's view on D.H. 9 A. The L.B. camera on mount. The Aperiodic
Centesimal compass. The strut thermometer . . . . ,, 112

The wind gauge bearing plate . . . . ,, 157

DIAGRAMS AS INSETS

Nos. 1–15. Analysis of experimental flights . . . . . . BETWEEN 40 & 41

No. 16.   Navigational survey of Eastern Counties (unadjusted) . . FACE 102

No. 17.   Navigational survey of Eastern Counties after elastic adjustment . ,, 103

* Photographic mosaic of district around Cambridge. Scale 1 mile = 1·53 inches.
* Error diagram of mosaic map.
* Oblique stereoscopic photographs.

*Available at www.cambridge.org/9781107511514

CHAPTER I

INTRODUCTION

THIS BOOK has been written as the result of a series of experiments made at Cambridge, by the authors, upon certain problems which arise in the construction of maps from aerial photographs. Its main purpose is to record and describe these experiments and to discuss the ways in which they may affect the choice of methods and equipment in aerial surveying; particularly that class of aerial surveying which is suitable for the mapping of large areas of imperfectly developed country, such as are still to be found in many parts of the world. It is not a comprehensive text-book upon the subject of aerial surveying; it does not, for instance, deal with the technique of photography, as applied to this purpose, nor with construction of finished maps from photographic data, and no attempt is made to enter into the details of methods and processes other than those directly related to the experiments under discussion.

Although our primary object is, as stated, to record and draw inferences from specific experiments, this and the next two chapters will be devoted to a discussion, in general terms, of the whole question of aerial surveying. This general discussion, which is as brief as we can make it, is necessary for a proper understanding of the various experiments which are to be studied in later chapters and for a full appreciation of the bearing of their results upon the practice of aerial surveying. The reader is advised not to skip these chapters, for later chapters have been written on the assumption that they have been carefully studied.

It is hoped that the book will be of interest to others besides the technical expert accustomed to mathematical processes, and therefore detailed discussions, involving anything but the simplest mathematical symbolism, have been banished to a series of appendices, of which the conclusions only have been brought into the main text. These appendices are logically necessary to support the arguments advanced, but they need not be read by anyone who is prepared to take their conclusions for granted. An exception must, however, be made of Appendix I in which the methods used for dealing with errors and for specifying the accuracy of a set of observations are explained. This appendix is short and is written in a form which we hope will make it easily understood by the non-mathematical reader. It might with advantage be studied by all who are interested in anything more than the mere conclusions drawn from the experiments, for the methods that we have used, although they were developed many years ago by Sir Francis Galton, differ somewhat from the conventional methods of studying errors which are at present in general use. For the reader who is interested only in conclusions and does not want to concern himself with experimental details, chapters IV and VII may well be omitted altogether.

It is a matter of common knowledge that photographs taken from aeroplanes were extensively used in the recent war, to assist in the construction of maps of enemy territory, and it is no secret that extremely useful maps were produced in this manner, containing an immense wealth of detail which could not have been recorded in any other way. Knowing this, it may be natural to assume that the difficulties inherent in the construction of maps from aerial photographs had been overcome and that the method could have been applied to the mapping of the world in peace, immediately on the cessation of the war. Such an assumption would not be correct, because there are certain factors which enter into the problem of mapping in peace which did not enter into the majority of the corresponding problems solved during the war.

The first of these is the economic factor. In the war the information required could be obtained in no other way, and was almost beyond price; in peace it is, generally speaking, true to say that the aeroplane will not be used at all unless it provides information of greater value for money spent than any alternative method; and in nearly every case there will be other more thoroughly tested methods available which will compete for the work. The second factor is that the bulk of the war surveying dealt with country for which maps, in some form, were already available; all that was required was to fit the aerial photographs to these maps, with the object of recording additional detail, or changes in detail. Much of the peace mapping of the world, on the other hand, has to be carried out in country for which no previous maps exist and it will not, in general, be economically possible to make a close ground survey to act as a basis for the aerial map, because this amounts to carrying out the survey twice over, a procedure only justifiable in special cases.

It will appear, as we proceed, that these two factors which differentiate aerial surveying in peace from the bulk of aerial surveying in the war, are of such great importance that they entirely alter the order of difficulty of the problem. It is sufficient, for the moment, to note that this is the case without going into detailed reasons, and to state that very much more attention to the specialised training of the personnel, and to the provision of suitable equipment, is necessary for commercial success in peace than was required to produce the valuable results that became an almost commonplace feature of the war.

In one theatre of war, Palestine, mapping from aerial photographs was carried on under conditions much more nearly akin to those that will obtain in peace than was the case in any of the other fronts. Here, the area to be surveyed was relatively very great and the material equipment strictly limited, so that the economic factor—the cost in material and personnel per square mile—was of great importance, whilst the almost complete absence of detailed maps made it necessary to construct the aerial map with the minimum of ground control. In the face of these difficulties successful aerial mapping was carried out on this front, and much experience was obtained which has a direct bearing upon the problems of economical aerial surveying in peace.

At the close of the war one of the officers who had been directly connected with this aerial mapping on the Palestine front had become convinced, as the result of his experience, of the great future of the aerial method of surveying, provided that correct methods and carefully trained personnel were employed; he was, however, equally convinced that any attempt to carry out aerial surveys on a large scale, without specially trained men and carefully thought out methods, would be a disastrous failure, at least from the commercial point of view. The latter belief was considerably strengthened by the failure of an attempt to carry out an experimental survey in India immediately after the war.

This officer, Captain H. Hamshaw Thomas, suggested the initiation of the experiments described in this book, and throughout has taken an active interest in them, his advice and help being invaluable. These experiments were, in fact, undertaken after considerable discussion, with Captain Thomas, of the methods used in Palestine and the difficulties encountered in the course of that survey. It appeared from these discussions that the success of the Palestine Survey was mainly dependent upon very accurate flying, carried out by crews of pilots and observers who were kept continually on the same work, at which they eventually became very expert. The earlier experiments at Cambridge were therefore arranged to test, in a scientific manner, the accuracy of flying which it is reasonable to expect under such conditions, and later experiments were designed to find the quality of the maps which would result from flying of this accuracy.

Maps, as everyone knows, are required for a wide variety of purposes, and the methods of surveying must, therefore, vary widely, in accordance with the type of map which it is desired to produce. It is thus obvious that the value of any particular method of mapping, such as that of using aerial photographs, will depend upon the type of map which is desired, and it is further obvious that several different schemes of aerial mapping may be required, each scheme being suitable to the construction

of its own particular type of map. Take, for instance, the earlier maps of
entirely unexplored country; here a series of photographs, showing up
the nature of the ground and its main features, might be of the greatest
value, even although they could not be compounded into anything but
a rough map, subject to considerable geometric errors. Consider, on
the other hand, the maps required for legal purposes in well-developed
country; here geometric accuracy may be of paramount importance and
a series of aerial photographs might have little value unless they could
be compounded into a very accurate map. Methods which might be of
great value in the first case might be useless in the second. It is perhaps
not quite so obvious that the reverse may be the case, since a map can
never take any harm from being too accurate, but even this becomes
obvious the moment the economic factor enters into the problem, for
it will not, in general, be possible to spend anything like so much money,
per unit area, upon maps of the first class as upon those of the second.
These are extreme cases, but they will serve to show that there is no
single answer to the question of the extent to which aerial methods
can replace the older ground methods. It is even unlikely that any one
aerial method will be adequate to deal with all those types of surveying
in which aerial co-operation is found to be of value.

The outstanding features which distinguish the aerial method of
mapping may be stated briefly as follows: the maps, being made from
photographs, can include an immense amount of detail, far more than
can be recorded as the result of any but the most expensive form of
non-photographic surveying. The photographs, being taken from a
commanding position, can give a simultaneous view of points which,
from the ground, could not be related without expensive clearing of
trees and other objects. Finally, the high speed of flight, coupled with
complete independence of obstacles that impede progress on the ground,
enables the aerial part of the survey to be carried out at enormous speed.

These features may be definitely set down as advantages, but they
are advantages which differ in value with the nature of the work in hand.
In contrast to these advantages, the aerial method, as at present understood, suffers from certain disadvantages. It is not, for instance, always
possible to obtain by it the high degree of accuracy which almost automatically is realised in the majority of ground surveys. This is particularly the case in those types of aerial surveying which lend themselves to
rapid and economical working, with a minimum of assistance from the
ground; that is to say, in just those cases where the economic factor is
likely to be strongly in favour of the aerial method. Again, there may
be serious economic difficulties in providing suitable landing grounds for
the aeroplanes; this is a factor which requires most careful consideration

in connection with every proposition for aerial co-operation in map making; it may often be the determining factor which alone prevents the aerial methods from being employed; it is a factor, however, which can only be dealt with in relation to each proposition separately, and it is not within the province of this book to deal with it in more than very general terms.

It should now be clear that the advisability of employing aerial methods in any specific survey will depend upon the relative importance attached to the three following factors:

1. Geometric Accuracy.
2. Economy, of Time and Material.
3. Detail.

Stress laid upon Geometric Accuracy will generally, but not always, react against the aerial method. Stress laid on Detail will always react in favour of the aerial method. Stress laid on Economy, may, or may not, favour the aerial method, according to the nature of the ground to be covered: for example, the obstruction offered to ground travel and visibility may be set against the difficulty of constructing aerodromes and providing for forced landings. It would seem, from these considerations, that it is useless to attempt to decide for or against the use of aeroplanes in connection with any particular survey, until a decision has been made upon the relative importance of these three factors in the problem under consideration. After deciding this question of relative importance, the final decision, for or against the aerial method, may be easy or it may present considerable difficulty.

As an example of an instance in which the decision should be easy, consider the survey of a flat, wooded country, interspaced with clearings or stretches of water suitable for landing an aeroplane; here, if ever, the aeroplane will be of great value. Consider, on the other hand, a geodesic survey, to determine accurately the relative positions of points which are very far apart. Here it is unlikely that the aerial photograph will have any value. The majority of surveying problems will lie between these extremes, and in these the decision between ground and aerial methods, or the decision as to the relative proportions in which ground and aerial methods are to be used, will require an exact knowledge of the accuracy to be expected from the aerial method, at various rates of expenditure per unit area. One of the objects of this book is to provide data from which this information as to accuracy and expenditure can be estimated, but it will be noticed, on reading the book, that no direct reference is made to costs in the form of money; this is because the cost will vary greatly with local conditions. It is thought, however, that the information given upon the equipment required and the number of hours flying per unit area should be sufficient to allow the costs of a specific survey to be estimated with reasonable accuracy.

It will be seen, as the book proceeds, that the possible methods of aerial surveying can be sharply divided into two groups, one of which will give accurate results, but is not very economical, whilst the other may be very economical, but is not, as yet, capable of giving very high accuracy. The methods of the first group, which have been intensively studied in Germany, will be of value in connection with surveys in which the factors of accuracy and detail are of great importance as compared with the factor of economy: for example, the survey of a town, in which the value of the map per unit area is high and accuracy and great detail are essential. The methods of the second group are those with which this book is concerned; they will be of value in connection with surveys in which the factors of economy and detail are of great importance compared with the factor of accuracy: for example, the survey of large areas of undeveloped country.

The methods to be discussed will thus be such as lend themselves to rapid economical working with great detail, but they will not, when carried out in their most economical form, give very great accuracy. It should be noticed however that the final accuracy reached will depend upon the assistance obtained from ground methods; if a close network of ground-survey control points can be provided, the rapid methods can be made to give considerable accuracy, but the provision of such a close network, in addition to the aerial part of the work, will take away from the very factors of economy and rapidity which may be the primary reasons for adopting aerial methods. It is important, however, to note that the same photographs, which have been used in the first place to produce a moderately accurate map with very little assistance from ground surveying, can be used again to make a more accurate map when additional ground control becomes available. We shall refer to this point again in the last chapter, when the practical applications of these methods are under discussion.

CHAPTER II

THE PROBLEM OF PHOTOGRAPHING THE WHOLE GROUND

IN all forms of aerial surveying which are required to give anything more than a series of disconnected views of the ground, two problems have to be faced: it is necessary

(1) To cover the ground with photographs which leave no gaps between them;

(2) To construct a true map from the photographs.

The fact that the second problem involves serious difficulties appears to be generally recognised, but the existence of the first problem is often entirely overlooked, or its difficulties underrated. It is, nevertheless, a mere truism to say that so long as the first problem remains unsolved the survey will be a failure, however complete the arrangements for dealing with the second problem. One cannot construct a map from photographs which do not exist. This chapter will deal in a general way with the first problem, whilst chapter III will be occupied with the second problem. In both these chapters the fundamental principles that underlie these problems will be treated in a general way, without reference to specific experiments which will be discussed in detail in later chapters.

Let us then turn to the problem of covering the ground without leaving gaps. The order of difficulty of this problem will depend very much upon the economic factor and upon the nature of the ground to be mapped. For example, in the survey of a town the area will be small and the value of the map per unit area will be, relatively, very great, so that the expense of taking many more photographs than are strictly necessary, or even the expense of making additional flights to fill in gaps, may be of minor importance. Or again, if the area is already well mapped, so that the aerial photographs are only needed to provide additional detail, the difficulty experienced by the pilot in locating the ground to be photographed will not be serious. If now we go to the other end of the scale and consider large tracts of undeveloped country, which have not hitherto been mapped in any detail, the difficulties connected with the problems we are discussing are at their worst. It will now, in general, be necessary to cut down the expense per unit area to a very low figure; this will involve the reduction, to an absolute minimum, of all waste due to excessive overlapping between the photographs and to unnecessary flying, but the difficulty of doing this will be greatly increased by the lack of accurately located features, which can be easily identified from the air. It has been already stated that we shall be concerned with problems which approximate more nearly to the latter of the two extremes above quoted than to the former, and it is therefore essential for us to make a particularly thorough study of this problem of covering the ground completely in the most economical manner possible.

The difficulties associated with this problem fall naturally into two groups:

(1) Those connected with the flying itself.

(2) Those connected with the identification of the ground to be photographed.

These two groups are to some extent interconnected, but we shall begin by dealing with the first of them. A moment's consideration will convince anyone that the most economical way of covering large areas of ground is to do so by a series of straight parallel flights, during which the photographs are taken at suitable intervals, so as to overlap fore-and-aft. These parallel flights must then be spaced at the right distance apart, to ensure a reasonable lateral overlap between the strips of photographs. Such a procedure will certainly be necessary when mapping any large area, whether the work be done by "vertical"* or "oblique" photographs.

For reasons of economy it is necessary to have as complete a control as possible upon the amount of the overlap, both fore-and-aft and lateral. The fore-and-aft overlap is easily controlled, provided that the speed of flight can be kept moderately constant and the camera axis approximately parallel to a fixed direction, for it is then only necessary to make the exposures at calculated equal intervals of time. The control of the lateral overlap, on the other hand, is a much more difficult matter, since it depends upon keeping all the flights straight, parallel, and spaced the correct distance apart, as well as upon keeping the camera axis parallel to a fixed direction. Now it will appear shortly that it is a matter of extreme difficulty to maintain the camera axis parallel to a fixed direction, unless the aeroplane is flying straight and at a constant speed; it will therefore be obvious that, both directly and indirectly, the economical covering of the ground depends upon the power to fly straight and at a constant speed.

It is found by experience that when pilots are asked to fly straight and steadily and nothing else they can do so with surprising accuracy,

* A "vertical" photograph is one in which the axis of the camera is intended to be vertical, but may be accidentally inclined at a small angle. An "oblique" photograph is one which is purposely taken with the axis of the camera inclined at a considerable angle to the vertical.

after very little practice; but it is unfortunately necessary for the complete solution of our problem that the successive flights, besides being straight in themselves, should be parallel with one another and have the correct lateral spacing. Now although it is relatively easy to fly straight and steadily when the particular track along which one has to fly is not specified, it is quite another thing to be asked to fly straight and steadily over a predetermined track on the ground, particularly when the track is not clearly marked. All experience, both of aerial surveying and aerial bombing, goes to show that this problem of flying straight over a predetermined track involves great difficulty, unless approached in the correct manner. It is this difficulty, more than any other, which has retarded the development of both aerial bombing and aerial surveying, and it is therefore advisable to reach a thorough understanding of the causes that underlie it, before proceeding further with our study of the subject.

It must first of all be realised that a pilot, situated at a height of, say, two miles above the earth, cannot fly accurately over an object on the ground unless he has a sufficiently exact appreciation of the vertical. If he wishes to pass exactly over a given object whilst flying on a straight course, he must fly so as to keep the object in a plane which contains the vertical and his direction of motion; or, in other words, so that the point vertically below his aeroplane—the plumb point as it is called—travels along a line which intersects the object. This is exactly what a pilot will naturally try to do when asked to carry out this operation, and he will have no great difficulty in doing it so long as he knows both the vertical and his direction of motion over the ground. If the position of the plumb point is accurately and continuously known, it is not difficult to make a reasonably good estimate of the direction of motion over the ground, but if this is not known and if the error in its estimation varies from time to time, it becomes very difficult to determine the direction of motion; hence we see that, both directly and indirectly, a knowledge of the direction of the vertical is essential to success in this operation.

Now it is a fact, familiar to all students of mechanics, that any horizontal acceleration impressed upon the pilot and his aeroplane will have the effect of altering his estimation of the vertical, in so far as it depends on the sense of balance, and what is still worse, the readings of all instruments designed to indicate the vertical will be altered in the same way, so that they support the erroneous evidence of his senses. There is, in fact, in that aeroplane, an apparent gravitational field which is not the same as the earth's field, and we shall call the direction of that field the "apparent vertical." We may thus state the proposition that the apparent vertical will only coincide with the true vertical when the aeroplane is subject to no horizontal accelerations; that is to say, when it is flying straight and at a steady forward speed. It is a fact, directly following from Newton's laws of motion, that no instrument either human or mechanical can be devised to distinguish the apparent from the true vertical, unless it depends for its action on something external to the aeroplane, such as the horizon or the undisturbed air surrounding the aeroplane.

There are certain apparent exceptions to the sweeping statement just made: it is possible to make a gyroscopic instrument which seeks the apparent vertical so slowly that it indicates the mean direction over a long period; if this period is long enough, the mean direction must approximate closely to the true vertical. Again, it is possible to use instruments which rely upon reactions from the air outside the aeroplane; these, in theory, are capable of indicating the true vertical independently of the accelerations of the aeroplane. Strenuous efforts were made during the war to use instruments of both these types, in connection with the problem of accurate bombing from aeroplanes, but little success was achieved, partly because of their very delicate nature and partly because of the difficulty of incorporating them in an aeroplane in such a way that the pilot could use them satisfactorily.

Rightly or wrongly, it was decided, before the experiments shortly to be considered were started, to avoid using the very delicate instruments that are required to distinguish between the true and apparent verticals in either of the above ways. It becomes therefore necessary to consider what other means of distinguishing the true vertical are at the pilot's disposal. Except when flying within a cloud, or on a dark cloudy night, there is always some rough indication of the whereabouts of the horizon, even on days when no exact determination of its position is possible, and when the position of the horizon is known the direction of the vertical is easily deduced. A pilot who can see enough of the horizon at one time is, as a matter of fact, able to keep his aeroplane on an even keel with a surprising accuracy, even when the weather conditions are such that a casual observer would scarcely believe that the position of the horizon is defined at all. The average pilot soon discovers that his sensations of the vertical are not reliable, and he rapidly finds that the only way in which he can manage his aeroplane successfully is by referring continually to the horizon, which not only tells him when he is on an even keel, but also tells him whether he is turning in azimuth. This knowledge may be gained consciously, but more often quite unconsciously, in which case the pilot never realises the extent to which he relies on the horizon, until he is deprived of it by entering a cloud or by trying to fly on a dark night, when it is not unusual for him to turn right over on his back before he discovers that he has departed from ordinary horizontal flight. It is an interesting fact that many of the early pilots were so unaware of the reason for their loss of control in clouds that they attributed it to such unlikely causes as the influence of atmospheric electricity upon their compasses, or even upon the aeroplane itself.

It will now be easy to realise that a continuous appreciation of the relation of the aeroplane to the horizon is the basis of all normal flying; from it the pilot can determine both the orientation of his aeroplane to the vertical and the rate of turn in azimuth, and in the absence of special instruments it is the only reliable source from which this information can be derived. At the height, and under the atmospheric conditions, suitable for aerial photography, the horizon is generally defined very clearly, so that it is easy to understand why the average pilot experiences no great difficulty in flying very straight and level when nothing else is required of him; for he can then sit up in his aeroplane and enjoy an uninterrupted view of a large sector of the horizon.

Suppose, however, that he is asked, in addition to flying straight, to cause his straight flight to pass over some definite object; he will have to take his attention off the horizon, both to locate the object and to make sure that he is going to pass over it; but we have seen that when his attention is off the horizon, his appreciation of the direction of the vertical may be in error. So far we have merely seen how an error may arise, but we have no information as to whether the error so introduced is likely to be serious or not; to throw light on this point let us now follow the course of events whilst the pilot is approaching the objective.

Suppose, for simplicity, that the pilot fixes his eye on the objective and does not look at the horizon at all; he will now have no guide to the direction of the vertical, other than his sensations, or the indications of instruments which can conveniently be placed in his field of view; these will all conspire to mislead him into the belief that the apparent vertical is the true one. Let us suppose that when he first looks down he is flying straight and steadily and, perceiving that he is going to pass to the left of his objective, uses his rudder to turn the aeroplane to the right. This operation has three separate effects, it turns the aeroplane to the right, it causes the left wing to rise and it rotates the apparent vertical so that the apparent plumb point moves to the left. It generally happens that the angle of roll of the aeroplane and the angle through which the apparent vertical is rotated are approximately equal, so that the apparent vertical is not altered relative to the aeroplane, and if the change is slow enough, the pilot is quite unable to appreciate that he is no longer flying on an even keel; he therefore continues to turn his aeroplane until the objective lies in a plane containing the direction of motion and the *apparent* vertical. Since the true vertical now lies to the right of the apparent vertical it is easy to see that the pilot will turn too far, and a more careful consideration will show that, if he continues henceforward to fly so that the objective remains in the apparent vertical plane, he will fly upon a curve which need not pass anywhere near the objective.

In fact, if the pilot maintains a constant height and flies always so that the objective lies in the plane which contains the direction of motion and the apparent vertical (that is if he flies so as to be entirely satisfied with what he is doing), then the aeroplane will travel upon one of a series of curves*, such as those in fig. 1, which show the paths that

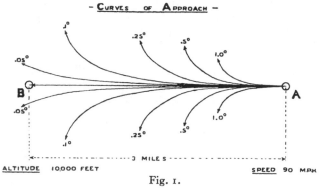

Fig. 1.

would be described by an aeroplane starting at *A* and attempting to fly over *B*, on the suppositions that *B* is kept always in the plane containing the apparent vertical and the direction of motion, and that the initial error of direction when at *A* is as shown on the curves. When an inexperienced pilot attempts to fly so as to pass over a definite object, he generally makes the mistake of looking almost exclusively at the object and paying very little attention to the horizon, with the result that he starts to fly along some such curve as those shown in fig. 1. If this happens to be one of the curves that passes reasonably near the objective, he may never realise that he has made any mistake at all, and he may return from his flight with a firm belief that he passed exactly over the required point; only to find, to his dismay and indignation, that independent records show him to have passed wide of the mark. At

* These curves have the common property $p\rho =$ constant, where p is the perpendicular from a point vertically above the objective upon a tangent to the track of the aeroplane at a point where the radius of curvature is ρ. The constant depends upon the height and speed of the aeroplane. This relation leads to a differential equation which is hard to solve mathematically, but which can be solved graphically by a step-by-step process.

other times it happens that the pilot starts out upon one of the curves which, if carried right through, would involve a large angle of roll in the aeroplane; in these cases he notices, sooner or later, that the horizon is no longer in its proper place; he then looks up, levels his aeroplane by the horizon and, looking down again at his objective, immediately starts off upon another of the curves of fig. 1. He thus approaches his objective in a series of curves, the curvature of which is often quite surprising.

These phenomena were thoroughly studied during the war, in connection with the problem of bombing. Great numbers of pilots were sent to fly over a camera obscura in which their tracks could be recorded, and it was found that nearly all inexperienced pilots made the mistakes which we have been discussing, but that it was possible to avoid these mistakes by training them to pay much more attention to the horizon than to the objective, during the approach. It was even found advisable, when an observer was carried, to divide the work so that the observer alone looked at the objective, whilst the pilot sat up straight in his aeroplane and attended to the horizon. Whichever of these systems is adopted the best procedure is found to be as follows. The pilot first levels his aeroplane and makes sure, by fixing his eye on a distant point on the horizon, that he is flying straight. Either he, or his observer, then looks down and estimates whether the direction of flight will carry them over the objective—he may or may not have some simple sight to assist him in his estimation. If it is considered that the aeroplane is not travelling in exactly the right direction, a small turn is made, but there must be no attempt to decide if the amount of turn is correct until the aeroplane has regained an even keel and is flying straight and steadily. There is not, in general, time for more than two or three such alterations of direction, hence the estimates must be accurately made in the first place.

We can now appreciate one of the numerous reasons why it is necessary to employ specially trained personnel, for it will obviously be impossible to carry out an economical survey without the ability to pass accurately over a predetermined point. We can also understand the importance of avoiding the use of methods which necessitate continual attention to the detail of the ground over which the aeroplane is flying; for if the pilot is to be continually repeating the difficult operation of flying over a given point he will be unlikely to fly with the straightness and steadiness required for economical work. The ideal procedure, in fact, would be one in which the pilot is not required to look at the ground at all during the time when the photographs are being taken; this implies that he should bring his aeroplane accurately over a given point at the start of the photographic flight and, henceforward, proceed in the required direction by compass, or other similar instrument, and without further reference to the ground. This is, in fact, the method used throughout our work.

The problem, at this stage, becomes complicated by the introduction of the second group of difficulties before mentioned, namely those relating to the identification of the ground to be covered. In highly developed countries such as England, in which there are a vast number of points that can be identified from the air and accurately located in maps, the difficulties that come under this head are not serious, and in these circumstances the following procedure might be adopted. Choose two or three points several miles apart on the required track and, taking the aeroplane well behind these, manœuvre so as to bring them into a line; then fly so as to keep them in a line and notice the compass bearing of the aeroplane which will do this, despite any cross wind that may be blowing. If the aeroplane then continues to fly at the same speed and compass bearing, it will continue along the correct track, provided that the wind does not alter. The operation of finding the compass course might be carried out before actually reaching the part of the track in which photographs are required, so that the pilot will not need to look down whilst the photography is in progress. Where it can be used this method might be quite a good one, but it depends on being able to find two or more points visible from the air and lying upon the required track, and is therefore entirely unsuited for work in previously unmapped countries, for in these there will be no means of telling the exact relative positions of points which form good landmarks from the air. Indeed the problem of the identification of correctly spaced tracks in previously unmapped country is one of considerable difficulty, which we will now proceed to study in general terms, leaving the detail consideration to a later chapter.

In the first place it is clear that it would be very unsound policy to rely upon artificially constructed landmarks. This would, we think, be a mistaken policy, even in highly developed country; in undeveloped country it would be obviously impracticable, both on account of the expense of sending out expeditions to place the marks, and because a fairly close preliminary survey would be required to place them with sufficient accuracy to be of any value. We have therefore to provide some scheme whereby the pilot can be guided into making a series of parallel flights with the correct lateral spacing, using only such landmarks as occur naturally. This can be done by the following method, which was used in the war surveys of Palestine. The survey is begun by the pilot flying from some starting-point, in a predetermined direction, keeping his speed and direction through the air as constant as possible.

During this flight vertical photographs are taken at equal intervals of time. These can be printed and formed into a strip showing a continuous view of an approximately straight strip of country, which can be roughly divided up into equal intervals, depending upon the time interval between the photographs, and the ground speed. It is now a simple matter to mark upon this strip suitable starting-points for the routine mapping flights, which may be at right angles to the first strip, and can be carried out in subsequent flights.

When a pilot sets out on one of his routine surveying flights he will carry one of these "Indication Strips," as we may call them, and use it as though it were a map, to locate the starting-points of his separate survey flights. If he succeeds in getting truly over these starting-points upon his indication strip, he will at least insure that his survey flights are correctly spaced at the start; if, in addition, he succeeds in making all his survey flights in exactly the same direction, they will obviously remain correctly spaced throughout their length. In practice, he will make finite errors, both at the starting-point and in the direction of each of his flights, so that gaps will begin to appear between the strips of photographs after a distance which will depend upon his errors and the lateral overlap allowed for in the first place. The length of the routine survey flights is, therefore, strictly limited by the accuracy of his flying and the overlap, or wastage, that is considered allowable.

After thorough consideration of this method and of several alternatives which had from time to time been suggested, we came to the conclusion that it is probably the simplest and best method to adopt in undeveloped country. It fulfils the condition that the crew of the aeroplane shall not be required to attend to the ground beneath them during the course of the actual photography, and it probably calls for the exercise of less special skill on the part of the pilot than any other method which will give good results. Even so, however, the operations involved in this procedure are by no means easy. A moment's consideration will show that it is not sufficient for the flights to be merely parallel; they must lie, approximately at least, in a predetermined direction, independent of any wind that may be blowing. This is because the survey, in general, will require many days for its completion, and it would be highly inconvenient if photographic strips taken on successive days were to be seriously out of parallel. Further, it will be necessary, on the score of economy, to take photographs on both the out and the return journey and, if these are to be parallel, an accurate allowance must be made for whatever cross wind is blowing at the time. Yet again, if the fore-and-aft overlap between the successive photographs of one flight is to be kept within economical limits, it is essential to know the approximate speed made good over the ground. All this implies an accurate knowledge of the wind at the height of the survey, and makes it necessary for the crew of the aeroplane to face all those problems that are involved in the accurate navigation of an aeroplane by dead reckoning. This in itself is no mean task and requires considerable specialised training and experience.

It should now be perfectly clear why the authors lay such great stress upon the necessity for employing none but specially trained and experienced crews, when a survey is required to be made in the economical manner which is essential for commercial success. It is not contended that it will be impossible for inexperienced crews to cover the ground completely, but that the wastage due to excessive overlapping, or to the necessity for additional flights to fill up gaps, will vary tremendously with the skill of the operators. Where, therefore, economy of time and material is an important factor in any surveying proposition, experience and training in the operators will be essential, merely to ensure that the ground is properly covered, to say nothing of the necessity for the accurate flying which we shall see later is required to enable the map to be compiled rapidly and economically, without excessive assistance from a detailed auxiliary ground survey.

It may be as well to mention at this stage that, although we hold strong views on the necessity for preliminary training, we do not consider that exceptional inherent skill is essential: on the contrary, we are of the opinion that, after the necessary training, any normally good pilot could achieve success in this work. Our ideas on the question of training are given in chapter VIII.

The above is merely an outline, in general terms, of the problems that must be faced before large areas of ground can be covered economically. The form taken by these problems in certain special cases will be considered in more detail in later chapters.

CHAPTER III

THE PROBLEM OF MAKING A MAP
FROM THE PHOTOGRAPHS

THE purpose of this chapter is to discuss, in general terms, the second of the two large groups into which the problems of aerial surveying were divided at the beginning of chapter II. We shall now consider what information is required to allow a true map to be constructed from a set of aerial photographs, on the assumption that these have been so taken that they cover the whole ground.

When the ground to be mapped is hilly and the hills are not small by comparison with the height of the camera, it is essential that every point to be mapped shall appear in at least two photographs taken from different points, and that the position and orientation of the camera at each exposure shall be known. The case of flat country, however, provides an exception to this general rule, of such great importance that it merits special consideration. Here, a series of photographs, taken from a constant height by a camera which has its axis vertical, can be combined into a true map of the ground, in the absence of any knowledge concerning either the position of the camera or its orientation in azimuth. To understand how this can be it is merely necessary to observe that each photograph forms a true plan of a portion of the ground and that the scale of all these plans is the same; hence the separate photographs, if they overlap at all, can be combined into one large plan by merely sliding them about until the detail of the ground runs continuously over the joins. We shall make a special study of this procedure in chapter V, but in the meantime merely note that this method can be applied successfully to country in which local differences of height do not greatly exceed one-twentieth of the height of the aeroplane and that, when it can be applied, it provides by far the simplest method of mapping with moderate accuracy and great detail.

With this short digression let us return to the general case in which it is necessary to determine the position and orientation of the camera at the moment of each exposure. The immense variety of methods by which this might be done can conveniently be divided into two groups: those that require the appearance upon each photograph of points which have been accurately located by independent means, and those in which this is unnecessary. The convenience of this division lies in the fact that the practical problems involved are entirely different in the two groups; thus the difficulties connected with the first group lie in the provision, by independent means, of an adequate network of accurately surveyed points, visible from the air, and in the reduction of the results in the office; while the difficulties connected with the second group of methods lie, for the most part, in the flying itself. We shall begin by dealing with the first group.

If three points, accurately known in position, occur in any photograph, the position and orientation of the camera at the moment of exposure can be determined completely by geometrical methods, all processes by which this can be done being given the generic name "re-section." If the points are very sharply defined on the photograph and very accurately located upon it, it is possible, by the process of re-section to determine the position and orientation of the camera with very great accuracy; complete systems of aerial surveying have indeed been developed entirely upon this principle.

The main advantages of the re-section method are that the accuracy of the resulting map may be relatively very great and is not directly dependent upon the accuracy of the flying, and, further, that it can be made to give quite accurate contours showing the height of the ground, a thing that is not easily done by any alternative aerial method that has yet been devised. The principal objection to methods that rely upon the re-section of each photograph is that an accurate auxiliary ground survey is required to provide the known points on which the re-section is based. Some control from auxiliary ground surveying is necessary in all systems of aerial surveying that have any pretence to accuracy, but the number of control points that are required per unit area is much greater when individual photographs have to be re-sected than when certain other methods are employed. Theoretically it would be possible to use re-section without the necessity of having independently located points appearing in each photograph, for a single pair of photographs might be re-sected from three points and used to provide additional points from which adjacent overlapping photographs could, in turn, be re-sected, and this process could be continued indefinitely. In practice, however, such a procedure involves serious cumulative errors, unless each operation can be carried out with very great accuracy, and we are not aware of any evidence which shows conclusively that it is possible to carry out a survey by the methods of re-section, without providing independently surveyed control points that will appear on each plate.

In some circumstances there may be no serious difficulty in providing a close and accurate ground survey on which to base the aerial survey; such a ground survey may, indeed, be in existence before the aerial survey is contemplated. In these circumstances there is now very little doubt that the re-section method, when properly applied, will give valuable results, but there will be many other circumstances in which the aerial method might be applied to regions for which no previous detailed survey exists, and in which the provision of such a close and accurate ground survey would either be impossible or would absorb all the funds available for the whole survey, and more. When, for instance, large areas of un-developed and possibly densely wooded country are in question, the value of the aerial method will depend mainly on the extent to which it can be carried on without much help from the ground. It then becomes necessary to employ methods that do not rely at all upon the re-section of individual photographs, methods in which the position and orientation of the camera is determined without the inclusion of accurately located points on each photograph.

From the foregoing considerations it seems probable that both classes of aerial surveying will be required, if full use is to be made of the aeroplane in the mapping of the world. For the early mapping of the large areas of the world that still await development, methods which do not depend upon re-section from known points in the photographs will be required, while the later and more accurate mapping of well popu-lated areas, or of small areas in which, for some reason, great accuracy is required, will almost certainly be carried out by the method of re-section. Very flat country may however afford an exception to this rule, for here the methods of chapter v may compete successfully with re-section, even when considerable accuracy is required. It does not follow that it will be necessary to carry out the aerial part of the mapping of a new country twice over, once for the earlier cheap map and later for the re-sected and relatively expensive map; on the contrary, one set of photographs could be made to serve both purposes, provided that the camera used in the first place fulfilled certain requirements of accuracy of internal adjustment.

The experiments on aerial surveying which have been carried out at Cambridge have been exclusively directed to the development of methods in which re-section is not employed, and the remainder of this book, which describes and discusses these experiments, will therefore deal exclusively with these methods. From what has been said, however, it should be clear that this exclusion of methods that involve re-section from the experiments and discussions does not imply the opinion that such methods have no practical value. The decision to restrict the Cambridge experiments in this way was come to merely as a matter of policy, after carefully considering the whole situation, in consultation with the Air Survey Committee—a joint committee of the three fighting services for the co-ordination and initiation of experimental work on this subject.

The experimental study of the group of methods that involve re-section has been carried on by this committee and by private individuals in this country, but has reached its highest development in Germany, where a remarkable apparatus, called in England the Auto-Cartograph, has been developed. Using this instrument it is possible to draw out an accurate map containing height contours, from direct observation of the photographs, which are viewed stereoscopically in pairs. This apparatus is very costly and cumbersome, but it appears to work exceedingly well and to remove what was hitherto one of the chief objections to the use of re-section, namely the immense amount of office labour that is required in the absence of some such elaborate instrument. The great expense and bulk of this instrument does not necessarily prohibit its use in connection with small schemes of surveying, for it seems reasonable to suppose that a single machine could be installed in some central position in each country and the photographs brought to it and reduced by a permanent staff. It will be seen therefore that the group of methods which depend upon re-section have by no means been neglected, either in this country or elsewhere, and that a considerable quantity of published information on the theory and practice of these methods is now available.

In excluding from our further discussion methods that involve re-section from known points on the photographs, we are not, of course, abandoning all help from ground surveying. On the contrary, we consider that all methods of aerial surveying will require some control from the ground to prevent the accumulation of errors over great distances. The distinctive feature of the methods we shall now discuss is, however, that the spacing of the points which will have to be located on the ground, or the "control" points, as we shall in future call them, will be very much wider than the distance covered in a single photograph. It does not of course follow that we advocate the use, in all cases, of a very open ground control—it may often be advantageous to use a close control, even in connection with the methods that we have studied. But it is, we think, important to develop methods that can, if necessary, be used with a minimum of ground control, since it will often happen that the circumstances most favourable to the use of the aerial method are exactly those in which ground surveying is most difficult; moreover, the main reason for adopting the methods to be considered, in opposition to the methods that depend on re-section, will be that there is some cogent objection to making a really close ground survey.

We shall now pass in review the methods that might be used to find the position and orientation of the camera at the moment of exposure, when the method of re-section from known points in each photograph is excluded. The camera has six degrees of freedom or, to state the matter otherwise, six independent quantities are required to specify its position and orientation. For the purpose in hand we may classify these six degrees of freedom as follows:

Tilt of the axis from the vertical	involving 2 degrees of freedom.
Orientation of the camera about the vertical, or azimuth of camera	involving 1 degree of freedom.
Position of camera in plan	involving 2 degrees of freedom.
Height of camera	involving 1 degree of freedom.

The object of dividing the six degrees of freedom into four groups, as above, is that the measurements required in connection with each of these groups are quite different in kind; thus the methods of measuring height are in general totally different from the methods adopted to determine position in plan, and so forth. When surveying flat country, for instance, it will be remembered that it may be unnecessary to concern ourselves with two of the groups, for we need not know either the azimuth or the plan position of the camera, provided that we keep the axis vertical and the height constant. We shall now proceed to discuss the four groups separately and in the above order.

The tilt of the optical axis of the camera from the vertical can be determined, either by direct reference to the gravitational field or, indirectly, by reference to the horizon. Direct reference to the gravitational field is, as we have seen, complicated by the fact that the apparent gravitational field within the aeroplane does not coincide with the true field, unless the aeroplane is flying straight and steadily. It follows, therefore, that the accuracy with which the tilt of the camera can be determined by simple instruments, such as the pendulum or spirit level, or by the sense of balance of the pilot, will depend upon the accuracy with which the aeroplane can be flown in a straight line at a constant speed. If this could be done with absolute accuracy, then there would be no more difficulty in setting the camera axis at any required angle to the vertical than would be experienced on the ground; if, on the other hand, the aeroplane is either turning or changing its speed, the problem becomes one of extraordinary difficulty.

It is clear that the determination of the tilt of the camera axis from the vertical is closely bound up with that same problem of flying straight and steadily which interested us when considering the difficulties of covering the whole ground in an economical manner; we now see therefore that the development of methods which will result in very straight and steady flying will go a long way towards the simultaneous solution of two of the most difficult problems in aerial surveying. We have already noticed the fact that it is not difficult to train a good pilot to fly straight and steadily, when he is not asked to attend closely to the ground immediately beneath him; but that it is very difficult for him to keep even approximately straight, if he is continually required to cause his aeroplane to pass over predetermined and possibly badly defined objects on the ground. If, therefore, the orientation of the camera is to be determined by direct reference to the vertical, there are two distinct reasons why it is important to arrange the procedure of a survey in such a way that the pilot is not required to attend to his position over the ground whilst the photographs are being taken, and to make sure that his training will be such as to develop, to the highest degree, his powers of flying straight and steadily.

If attention is not given to these points the result is almost certain to be a series of photographs taken with unknown tilts up to six degrees, or more, and covering strongly curved tracks which will, in some parts, overlap each other to a wasteful extent and, in others, will leave wide gaps entirely unphotographed. Such, for instance, was the result of the hasty attempt to carry out a test aerial survey in India during 1919. On the other hand, we have found that, by attending to the points above enumerated, there is no great difficulty in reducing the unknown tilt of the camera axis to an average value of about one degree—the error rarely exceeding two degrees—and that the flights themselves can be kept surprisingly straight over great distances. The evidence on which these statements are founded is given in chapter IV, but in the meantime it is sufficient to note that the improvement in accuracy that can be obtained by attention to the above points will, in many cases, make all the difference between striking success and total failure in an aerial survey carried out on a commercial basis.

It may be remembered that reference was made, in chapter II, to certain delicate instruments which can, in theory at least, be made to indicate the vertical independently of short period accelerations occurring in the aeroplane. Certain forms of these instruments could conceivably be used, either to hold the camera axis vertical, or to indicate the tilt of the axis upon the plate. We have already stated that it was our policy to push on with the surveying experiments without waiting for the development of apparatus of this class, and this is what we have done. Our opinion however is that it is unlikely that success will be achieved in the near future with instruments designed to hold the camera itself with its axis vertical, but that the alternative procedure of indicating the tilt of the camera axis on the photograph is more likely to be successful and may, in fact, lead to developments of great value. Even should the latter procedure be adopted, the pilot still may not be relieved from the necessity of flying as straight and steadily as possible, because the types of instrument that can be foreseen at present will not give an absolute indication of the vertical, but will merely cut down the excursions of the apparent vertical by some proportion, say a half or a third; so that the final accuracy of the result will still depend upon the accuracy of the flying. In any case, it is, as we have seen in chapter II, essential to be able to fly very straight and steadily if the ground is to be covered at all economically, and this necessity will not be avoided by any foreseeable developments of instruments that indicate the vertical. It follows, therefore, that it would only pay to use such instruments when it is important to determine the tilt of the camera with an accuracy greater than the variations of the apparent vertical in good straight flying, that is, with an accuracy greater than one, or at most two, degrees.

We have yet to consider the alternative method of determining the relation of the camera axis to the true vertical by reference to the horizon. In general, the ground or sea horizon is invisible from an aeroplane at a considerable altitude, the distance being so great that haze generally intervenes; but if the aeroplane is flying at a sufficient height, the haze itself nearly always provides a well-marked horizon that can be used to determine the vertical. A long series of experiments has recently been carried out by the Air Ministry upon the accuracy with which these haze horizons can be used to determine the vertical*. On eight different days in which haze horizons were visible without much cloud, the maximum error in the determination of the vertical from these horizons was found to be twenty-three minutes of arc, the average error being about ten minutes. These experiments were extended to cover the case of definite cloud horizons as seen from an aeroplane above them, the errors of which were found to be of the same order of magnitude as for the haze horizons. These errors are very much less than the accidental tilts that occur, even with the most skilful pilots, so that when the horizon can conveniently be used, it should give valuable help in any aerial survey in which an accurate knowledge of tilt is desirable.

In some countries it may happen that the air is so clear that the ground horizon itself can be seen, and then it might be thought that large errors would be introduced by irregularity in the ground. When the aeroplane is at a height of some ten thousand feet—a convenient height for mapping—the ground horizon is about a hundred miles away and, at this distance, a thousand feet subtends an angle of only six and a half minutes; the estimated vertical would therefore be in error by six and a half minutes for every thousand feet of error in the estimated difference of height between the aeroplane and the ground at the horizon. This simple calculation is merely included to emphasise the fact that the use of the horizon in aerial surveying is not dependent upon the

* Aeronautical Research Committee Report, Met. 105 (unpublished).

presence of haze. As a matter of fact, the ground horizon in England is rarely, if ever, visible from a height of ten thousand feet, and we have no information upon the frequency with which it is visible in other countries. A particular instance in which the horizon might be used with advantage in connection with a scheme of aerial surveying is discussed in some detail in chapter VI.

Let us now consider methods of determining the orientation of the camera in azimuth. If this is not to depend upon the inclusion within the photographs of independently located points, it must depend either upon the compass, or upon observations of the sun. Both these methods involve serious practical difficulty, where great accuracy is required. The compass suffers from three distinct defects:

(1) It is disturbed by the magnetic properties of the aeroplane.

(2) Its indications are, in general, erroneous whenever the apparent vertical in the aeroplane differs from the true vertical.

(3) When the aeroplane turns, a swirl, or rotation, is set up in the liquid in which the compass is generally immersed, and this swirl disturbs the compass during the turn and for some time after the turn has ceased.

The first defect can be partially overcome by careful compensation and the use of a curve showing the deviations on any course, but it is the cause of considerable trouble, particularly in aeroplanes with metal bodies. If metal aeroplanes are to be used for aerial surveying great care must, therefore, be taken to mount the compass so as to be as free as possible from disturbing magnetic influences.

The second defect cannot be eradicated in any way, unless the true vertical can be determined. The truth of this very sweeping statement is easily realised, when it is pointed out that the direction of the magnetic meridian is the projection, upon the horizontal plane, of the direction of the earth's magnetic field; unless, therefore, the true horizontal plane and the direction of the magnetic field are both known, it is impossible to define the magnetic meridian. The types of compass in general use are mounted upon a single point, slightly above their centre of gravity, so that the card is kept in a horizontal plane by the direct action of gravity. These compasses will, therefore, indicate the projection of the direction of the magnetic field upon the *apparent* horizontal plane and will not indicate the true magnetic meridian, except in special cases. Within the aeroplane there will thus be, for compasses of this type, an apparent magnetic North, which will deviate from the true magnetic North whenever the apparent vertical deviates from the true vertical.

If the pilot were able to hold the aeroplane itself in a fixed relation to the horizon, so that the vertical is thereby determined without direct reference to the gravitational field, a compass which would be unaffected by acceleration could be constructed, either by mounting the card upon a fixed vertical axis, or by using a "dip" needle pointing along the magnetic lines and mounted within a spherical bowl which carries meridian lines upon its inner surface. Such devices fail completely unless the pilot can keep his aeroplane in a fixed relation to the horizon, and it is probably better to adhere to the more usual forms of compass, which do not depend upon the aeroplane remaining upon an even keel, but which do depend upon the absence of horizontal accelerations. It is worth noting, however, that any device which succeeds in accurately defining the true vertical can also be used to overcome this particular form of compass trouble.

Whatever may be the final solution of the compass difficulty, the fact remains that all existing compasses in service are affected by acceleration and seek the apparent magnetic North, instead of the real magnetic North. All such compasses therefore possess certain peculiarities which should be thoroughly understood by any pilot, or observer, who intends to use them for more than a rough indication of direction.

It is easily shown that the deflection of the apparent magnetic North from the real magnetic North can be approximately represented by the expression $\alpha/g \tan \delta$, where α is the component of the aeroplane's horizontal acceleration perpendicular to the magnetic meridian, δ is the dip of the magnetic field, and g is the acceleration of gravity. We note from this expression that the deflection of the compass in any given manœuvre will be zero on the magnetic equator and increase progressively towards the magnetic poles, at which points it becomes quite useless. In the latitude of the British Isles the difficulties due to this cause are serious but not by any means insuperable.

The quantity α/g is, to a first order, the tilt of the apparent vertical, and $\tan \delta$, in the neighbourhood of the British Isles, is a little more than two, so that, in these regions, the apparent magnetic North will wander from the real magnetic North by more than twice the angle by which the apparent vertical wanders from the true vertical. Now we have seen that, even with the best flying, the apparent vertical will occasionally wander through two degrees from the true vertical, so that the compass must be expected, in these latitudes, to fluctuate occasionally, even with the best flying, over a range of about eight degrees; with inferior flying, or in bumpy weather, it must fluctuate still farther. We have here yet another reason why good flying and still air are essential to success in any but the simplest work.

Again we notice that it is accelerations perpendicular to the magnetic meridian that disturb the compass; these are produced by changes of direction when flying along the meridian and by changes of speed when flying across it. When flying towards the nearer pole, the effect of a turn, either to right or left, is to turn the compass card in the same direction as the aeroplane, so that the turn registered by the compass is reduced, or in some cases even reversed. In these circumstances steering by compass alone becomes very difficult. When flying away from the nearer pole the compass turns in the opposite direction to the aeroplane, and steering is actually facilitated by the error. In high latitudes therefore it is always easier to carry out a straight flight by compass when flying away from the pole than when flying towards it. In practice the aeroplane is not kept upon a straight course entirely by compass; other devices, such as a sun shadow within the aeroplane, or the view of a distant object near the horizon, are used to keep straight; the compass merely shows that the straight flight is in the desired direction. These defects do not, therefore, as might be supposed, render accurate navigation impossible, even in high latitudes. It is clear, however, that they should be thoroughly understood by anyone who intends to use the compass in an aeroplane.

The third source of error, that due to swirl in the liquid, also has a serious influence upon the way in which the compass must be used in a survey. After the aeroplane has been turning, the liquid in the bowl is found to be rotating, and this rotation, or swirl, does not die right away for something between one and two minutes, during which time the fluid reaction between the compass needle and the liquid deflects the needle and causes it to register wrongly.

One unfortunate result of the second and third of the above defects of the aerial compass, as it exists at present, is that no reliance can be placed upon its indications until after the aeroplane has been flying straight and steadily for a period that lies between one and two minutes. This is a very serious drawback, because in this time the aeroplane may have flown as much as three miles, and to waste three miles at the start of every survey flight would be to incur a very serious economic loss, particularly when the individual flights are as short as those described in chapter v. It is possible to avoid the worst consequences of this loss of time by using an additional gyroscopic instrument to which reference will be made, but the defects that we have been discussing appear to present a definite bar to the use in aeroplanes of present-day compasses for very accurate determinations of azimuth.

The alternative method of determining azimuth from the sun could always be employed whenever the sun is not too near the zenith, but it involves the use of special tables showing the sun's azimuth at any time, and to use these during a flight would add considerably to the troubles of an already heavily burdened observer. Difficulty may also be experienced in arranging a suitable instrument which will admit the sun's rays from any angle and transmit them onto a scale that can conveniently be observed by the pilot, and it should be noted that the determination of azimuth from the sun is dependent upon a knowledge of the horizontal plane, in exactly the same way as when the earth's magnetic field is used. The authors have very little experience in the use of the sun for the purpose of determining azimuth, but think that it might be used where compass troubles are particularly severe.

It will be realised, from what has been said, that the accurate determination of the azimuth of the camera is a matter of extreme difficulty, and that methods of surveying from the air which depend upon an accurate knowledge of the camera's azimuth are unlikely to succeed, unless that knowledge is obtained from internal evidence in the photographs themselves. In one very useful branch of aerial surveying, namely the surveying of flat country by vertical photographs, the azimuth of the camera is, very fortunately, not required, and in the other types of surveying which will be discussed, some means of obtaining azimuth indirectly is generally contemplated.

We shall now turn our attention to the third group into which we divided the camera's six degrees of freedom, and shall study the possible methods of locating the camera in plan. This location might be effected by means of observations upon the aeroplane by observers on the ground. Thus, two theodolites, continuously trained upon the aeroplane, would be sufficient to fix its position both as regards plan and height. It is quite possible that some such procedure might be adopted in countries where the air is very clear and the difficulty of obtaining suitable view points on the ground is not serious. For general purposes, however, this method has very serious disadvantages; it would be practically useless in flat wooded country, unless the theodolite stations could be raised quite clear of the trees, and it involves close co-operation between the ground and the aeroplane, a thing which is very difficult to bring about, except with the most perfect organisation. It is, however, a method which might be employed with advantage in mountainous country, but the authors have carried out no experiments upon it, nor are they aware of any published accounts of attempts to develop it.

If we rule out methods that require direct observation from the ground and methods that involve re-section from known points on the photographs, we are left with no means of locating position in plan except that of navigation by dead reckoning; for the remaining alternative of using astronomical observations taken from the aeroplane is at present incapable of giving the required accuracy. When only one point on the path of an aeroplane is known, it is not at present possible to locate the aeroplane accurately at other points on its path, since this would require a very accurate knowledge, not only of the speed and direction of flight through the air, but also of the velocity of the air itself, that is to say, of the wind at the height in question. It will be seen later that we are able to obtain a moderate estimate of the wind at the height of the aeroplane, but that this estimate is not accurate enough for the above purpose. When, however, the position of the aeroplane is known independently at two or more points of its flight, it is possible to locate its position at intermediate points with comparative accuracy, on the assumption that it has been travelling in a straight line and at a constant speed between the known points. The possibilities of developing a system of aerial surveying on this principle are investigated in chapter VI, so that there is no necessity, in the present chapter, to enter into details.

There remains to be considered the methods of determining the height of the aeroplane. Here again, if we are to use neither re-section nor observation from the ground, we are left with only one method, that of measuring the pressure of the air and deducing the height indirectly. The height that corresponds to any given change of air pressure is dependent upon the air temperatures at all intermediate heights, hence it is obvious that measurement of height involves measurements of pressure and temperature. Strictly, it is necessary to measure temperature at all intermediate heights, but results that are accurate enough for many purposes can be obtained by measuring temperature at the height of the aeroplane only, assuming an average rate of temperature change with height.

The accuracy of height measurement which would be obtained in this way, when using perfect instruments, has been very thoroughly investigated by the Air Ministry. The method of investigation was to adopt some standard rule for the calculation of height from simultaneous measurements of temperature and pressure at the height in question, and to check the result against calculations depending upon the actual temperature distribution which was found, by experiment, to occur on a great number of occasions throughout the air between the ground and the aeroplane. Working on these lines, it was shown that the height deduced from a single measurement of temperature and pressure need never be in error by more than one per cent. This result, which applies between 6,000 and 30,000 feet, assumes accurate instruments and correctly chosen formulae for the deduction of the height from the reading of the instruments. If the instruments themselves are inaccurate, the errors may be greater than those indicated above, and although the additional errors due to imperfect instruments may be appreciable when the ordinary service aneroid is used, they can be reduced, by the use of special aneroids, to a quantity that is negligible in comparison with the errors resulting from variations in the temperature gradient. This question will be considered in more detail in chapter VII.

In many circumstances, particularly when mapping flat ground by vertical photographs, it may be more important to keep the height of the aeroplane constant than to know its exact value. The problem of keeping at a constant height is partly personal and partly instrumental. The pilot has to fly so as to keep the indicator of his aneroid steady, and the aneroid must have a sufficiently open scale and be sufficiently sensitive for this purpose. The standard service aneroid does not quite fulfil these conditions and special aneroids of greater sensitivity and more open scale are required. A long series of experiments on this problem of maintaining a constant height are described in chapter IV.

In the present chapter and in the two preceding it we have examined certain general principles relating to our subject. We have done this so that we may not have to interrupt the discussion of our experiments by somewhat lengthy theoretical digressions. Experience has shown us that many of the more serious difficulties confronting the beginner in aerial surveying are of a kind not readily foreseen, even by men of great general experience in aeronautics. The fundamental principles from which these difficulties rise must, however, be thoroughly understood before it is possible to make an intelligent study of the method by which they can be overcome. Our own experiments having been directed towards the development of rapid economical methods, we formulated, for our guidance, certain maxims, the significance of which will be readily grasped by the reader who has studied these first three chapters. These maxims may be stated as follows:

(1) The methods to be studied must not depend absolutely upon the provision of a *close* net of control points, accurately surveyed by independent means. The wider the spacing between control points that can be used with a given method, the more valuable, for general purposes, will that method be.

(2) It should not be necessary to provide co-operation from the ground, either during, or before, the flight.

Thus, methods that involve observing the aeroplane from the ground and methods that require artificial landmarks are to be avoided.

(3) The success of the survey must not depend absolutely upon correct working of complicated apparatus, additional to the well-tried instruments that are standard at the time.

It will appear later that we do not imply by this that no special apparatus should be used. On the contrary we definitely recommend the use of certain highly specialised apparatus, which we consider should be carried on every aerial survey. The idea behind this maxim is that all the apparatus used in the air, on a survey, must be extraordinarily reliable if it is to be of any value, and that every bit of additional apparatus which is absolutely essential to success is an additional source of weakness, particularly when working in undeveloped country, far from sources of supply and well-equipped workshops. For this reason we think that, however well the expedition may be equipped with specialised apparatus, the training of the pilots and the methods used should be such that the survey could continue, with perhaps a considerable reduction in the area covered per day, after the specialised apparatus has broken down. It is probable that the force of this maxim will decrease as more experience is gained, but it may always apply to new apparatus of a delicate nature during the first few years of its use in the field.

(4) The value of any method will depend upon the time required for the manipulation of individual photographs and it is most important to keep this time down to a minimum. Where possible, contact prints straight from the negatives should be used directly, but if any manipulation, such as the determination of tilt, is necessary it should be of the simplest character.

The importance of this last point will be grasped when we take into consideration the immense number of photographs required to cover such large areas as the undeveloped portions of the British colonies.

It cannot be too strongly emphasised that these maxims apply only to the type of surveying under consideration, namely the survey of large areas of undeveloped country. In other circumstances they may not apply at all.

CHAPTER IV

PRELIMINARY EXPERIMENTS

THIS chapter will deal with a series of experiments upon the accuracy with which a camera can be carried in an aeroplane at a constant height above the ground and with its axis vertical. It was explained in chapter III that, if this can be done with sufficient accuracy, it is possible to map flat country from the air without any exact knowledge of the position of the camera in plan, or its orientation in azimuth. It was mainly this type of aerial mapping that we had in mind when the experiments to be discussed were started, but it was realised that the information obtained from the experiments would be of fundamental importance in relation to all types of aerial surveying which do not rely upon re-section from known points on the ground. For the maintenance of a constant height, although only strictly necessary in connection with the particular type of surveying above mentioned, is yet a matter of considerable interest in relation to any type of aerial surveying in which economy and rapidity of reduction of the photographs to the form of a map is a serious consideration. Again the power to hold the camera axis vertical is evidence of the power to hold it at any predetermined angle to the vertical. Even should gyroscopes be used to average the excursions of the apparent vertical, the accuracy with which any such instrument will perform its function depends directly upon the magnitude of these excursions, so that it will still be necessary to know how to control them.

It will be realised, from what has been said in preceding chapters, that the ability to hold a camera axis vertical depends either upon the pilot's ability to fly sufficiently straight and steadily to avoid serious disturbances to the apparent vertical, or upon his power to hold his aeroplane level by reference to the horizon. Hence it will be obvious that the experiments now under discussion are really an indirect test of skill in flying, and the results will be subject to a personal element, which will vary from pilot to pilot. Two pilots only were employed on the experiments, so that it might reasonably be argued that they may not be representative of the average skill. If reliance is placed only on the evidence contained in this chapter, the possibility of this cannot be denied, but the authors themselves are of the opinion that no serious difficulty will be experienced in getting, from the majority of experienced pilots, results comparable to those here recorded, provided that they are properly instructed and practiced in the operations required. This opinion is based, apart from personal flying experience, upon two independent experimental sources. In the first place a series of experiments upon straight, steady flying, for the purpose of taking sextant observations of the altitude of the sun, were carried on simultaneously, but quite independently, by the pilots engaged in the present work and by several pilots of the Air Ministry Research Department. The results obtained from these independent groups were, on the average, almost exactly the same. In the second place a prolonged experience of the experimental training of bombing pilots in the war led to the conclusion that the majority of good pilots, when properly trained, all reached about the same order of accuracy in their flying, and this accuracy was comparable with that deduced from the results here obtained. Finally there is evidence that a third pilot, who carried out some of the experiments in chapter VI, was capable of flying with an accuracy equal to that of the experiments now to be considered. This evidence is discussed in chapter VI.

Mention has already been made of the fact that it is very much more difficult to fly straight and steadily, when looking at the ground beneath, than when sitting up with a full view of the horizon. This was well known before the experiments were started, having been discovered in connection with experiments on aerial bombing during the war. It was decided, therefore, that the experiments under discussion should be carried out in a manner that would involve the minimum attention to the ground beneath, and so allow the pilot to obtain the best results of which he was capable; the problem of applying accurate flying to actual map making would then be studied separately. The account of the attack upon this problem occurs in chapters V and VI.

The experiments discussed in the present chapter were to consist of a series of flights in which every effort would be made to keep the aeroplane at a constant height and to hold the camera with its axis vertical; means had therefore to be devised for determining the accuracy with which this had been accomplished. From an experimental point of view it is not an easy matter, either to determine the exact height, or the exact tilt, of an object such as a camera in an aeroplane. It is obviously impossible to use pendulums or spirit levels to determine the tilt, since the erroneous indications of these instruments are the primary cause of the errors to be measured. Again, the haze-horizon cannot be used, since it is partly by means of the horizon that the pilot obtains his idea of the horizontality of the aeroplane; separate evidence of the horizontality of the haze-horizon would therefore be required to make the results conclusive. The sun could be used to give one component of the tilt of the camera axis, but it cannot alone give the tilt from the vertical, because a knowledge of the angle between the camera axis and the direction of the sun gives no more information than that the axis lies on a certain cone, and thus does not completely define its relation to the vertical, except when the sun is in the Zenith.

After a careful consideration of all the methods which had from time to time been suggested for determining the orientation of some object within an aeroplane in flight, it was eventually decided that the only one which would give the required certainty would be the method of re-section, applied to photographs taken by a camera in the aeroplane. Thus, although the idea of using re-section on a survey had been abandoned, the method was used to provide the data on which the possibility of making a reasonably accurate survey without it could be investigated.

The process of re-section requires the presence of large numbers of accurately surveyed points, visible upon a photograph taken from the height in question, and the calculations are simplified if all these points are in the same plane. It fortunately happens that there is, near Cambridge, a stretch of fen country, over twenty miles long and several miles wide, which is so level that the height variations in it seldom exceed 10 ft. This country is closely intersected by narrow drains, all marked on the six inch ordnance survey map and easily visible in photographs taken from 10,000 ft., the height at which we proposed to carry out the experiments. Down the entire length of this strip of land run two wide straight drains, or canals, called the Bedford Level; these form very pronounced landmarks and are visible from an aeroplane at a great distance. This strip of country was ideally suited to our purpose; the intersections of the small drains would provide the means whereby the re-section of the photographs could be effected and the long straight drains would provide very suitable landmarks, by which the pilots could fly along the strip with a minimum of attention to the ground beneath.

The form taken by the experiments was that the pilot flew along the canals at a height, by aneroid, of 10,000 ft., paying particular attention to keeping the height constant and the aeroplane level. During these flights the observer took a series of photographs, at approximately equal time intervals, from a camera fixed in the aeroplane. These photographs contained the known intersections of the small drains already mentioned, and, therefore, the position and orientation of the camera at the moment of exposure could be calculated.

In the earlier experiments the camera was held fixed in the aeroplane, so that its orientation determined the orientation of the aeroplane itself. In the later experiments a new mount was used that allowed the fore-and-aft tilt of the camera, relative to the aeroplane, to be altered by a slow motion screw, which could be adjusted in flight in accordance with the average indications of a small quick reading spirit level. This adjustment was generally made at the beginning of an experiment, and only occasionally altered if the indications of the bubble were seen to be permanently in error. The camera movements, therefore, still recorded temporary angular movements of the aeroplane, and this addition merely ensured that there would be no permanent fore-and-aft tilt of the camera, due to errors in its original adjustment or to permanent changes in the tilt of the aeroplane. It will be see that in the later surveying experiments the camera was always adjusted in this way during flight.

One object of the experiments was to determine how nearly a camera can be held with its axis vertical; and, therefore, in the earlier experiments, during which no adjustments in the air were possible, the camera was very carefully adjusted on the ground before the experimental flight. This was done in the following manner. The aeroplane first ascended and carried out a mock experiment, flying level at the appropriate speed, and the average fore-and-aft tilt of some part, such as a longeron, was measured by means of a spirit level. The aeroplane was then set up in its shed, with this part at the same tilt and with the wing tips level and, whilst in this position, the camera was adjusted with its axis vertical. The primary effect, therefore, of the provision of the fore-and-aft adjustment in the air was to dispense with the necessity for this preliminary experiment.

Throughout the experiments there was no possibility, whilst in the air, of adjusting the lateral tilt of the camera relative to the aeroplane, so that, in regard to lateral tilt, the tilt recorded by the camera was the actual tilt of the aeroplane itself.

The aeroplanes used in the experiments were of the type known as the D.H. 9 A. They may be described as stable tractor two-seater aeroplanes, with a single 400 H.P. engine. They are generally considered to be comparatively easy to fly steadily. A photograph of one of these aeroplanes faces this page, where it is shown propped up in its shed, in approximately flying position. The pilot sits in the cockpit on a level with the circular identification mark, whilst the observer sits just behind him. The camera and gyroscopic rudder control are still farther down the body, in the position indicated by the laced fabric panel, which can be removed for inspection.

In all but one of the recorded experiments, the instruments used were the standard instruments which are normally carried by this aeroplane on military duty. A description of these instruments, together with that of the camera and other apparatus, and a general account of the methods of flying and the routine followed in the air are given in chapter VII. The remainder of the present chapter will deal mainly with the results obtained and with a discussion of their interest in relation to aerial surveying in general.

The D.H. 9 A

The methods used in the re-section of the photographs depend upon the fact that the absolute value of the tilts to be measured are small, never exceeding four degrees, and that, when this is the case, the components of tilt, in any two directions, can be determined separately and combined vectorially to give the resultant tilt in magnitude and direction, with all the accuracy required. Reference points are chosen near the corners of each photograph, and the distances between these points and the corresponding points on the six inch ordnance survey map are compared and used to give the component tilts about fore-and-aft and transverse axes in the aeroplane. From these components the magnitude and direction of the actual tilt is calculated and, when this has been done, it is a very simple matter to determine height, position in plan, and azimuth orientation of the camera. In theory only three reference points are necessary for complete re-section, but in practice we used anything up to twelve separate points for the purpose, because, by so doing, the processes could be simplified and the accuracy increased.

The determinations of the height and of the components of tilt from the vertical were carried out at least twice for each photograph, using entirely independent points of reference. It will be seen therefore that, in all, we used as many as twenty or more points of reference on each plate. This was made possible by the large numbers of small, but accurately mapped, drains with which the country is covered. It was the normal practice to make two such independent calculations for each plate, but when serious discrepancies occurred a third independent calculation was made. Every separate determination is included in the records, so that the data establishes, amongst other things, the accuracy of the particular method of re-section employed. The data has been carefully analysed to find the 50 per cent.* error of a single determination of tilt, and it is found to be in the neighbourhood of one quarter of a degree. It is estimated that the greater part of this error is due to the difficulty of locating the same points on the photographs and maps, rather than to errors in the actual measurement of the distances between points when located; hence it would seem that this is about the accuracy with which a vertical photograph can be re-sected from natural marks recorded on the six inch ordnance survey map. It would seem that, if greater accuracy is required, artificial marks would be necessary, especially designed to be very accurately defined in a photograph.

A parallel study of the accuracy of the determination of height shows a 50 per cent. error of the mean of two observations of about ten feet, but, in this case, there are other sources of error, such as the unequal

---

* The 50 per cent. error is that error which is equally likely to exceed, or be exceeded by, any individual error. See Appendix I.

stretching of the maps and inaccurate joining of map sheets. These errors would be more or less constant for one photograph, whatever points of reference are used, so that, in the case of the height measurements, the absolute errors may be somewhat greater than would be indicated by the consistency of the data.

The processes involved in this method of re-section proved to be very laborious, the complete analysis of a single photograph, with checks, requiring, on the average, nearly a day's work for one man. A special computer was engaged for some time to cope with this work; but even then we did not analyse all the available data; enough however has been analysed to obtain the information that we were seeking. These facts will give some idea of the labour involved in working accurately by re-section, when no elaborate apparatus, such as the Auto-Cartograph, is available.

One advantage of the process of re-section was that it gave all the information possible about the position and orientation of the camera at the moment of exposure. The statement of the results thus involved the record of three co-ordinates of position and three angular components of the orientation of the camera. We thought it advisable to record all these six variables, although three only, namely height and the two components of tilt, were of direct interest in connection with the method of mapping by " vertical " photographs which was in our minds at the time. All the data has been grouped together in Diagrams 1 to 11, the first of which faces p. 40. These set out the results of eleven separate runs along the canals, each run containing as a rule 18 separate exposures. We shall explain the meaning of these diagrams before discussing in detail the results that they represent.

Each diagram refers to a separate run along the canals, some of which were carried out consecutively in a single flight. The top figure (A) of each diagram is a plan of the tract over which the aeroplane flew; the dots represent the plumb points*; on the same diagram the centre lines of the main Bedford Level canals, which were used as landmarks, are shown as thin lines. It will be observed that many of the tracks are anything but straight and it might be supposed that the curvature indicated would correspond to considerable tilts of the apparent vertical. The curvature, however, is accentuated by the small scale of the plan view, and it will be seen later that it was exceedingly rare for the general curvature to be sufficient to account for a tilt of the apparent vertical of as much as one degree. Even so, the flights are often too curved to be of much use in a large survey; they do not, however, show the degree of straightness that can be reached in a long flight; this is definitely proved

---

* Points vertically below the camera.

by the results of later experiments to be discussed in subsequent chapters. In the flights now under discussion the bulk of the pilot's attention was directed towards maintaining a constant height and a level aeroplane, and the straightness of the flight was only considered in relation to these requirements. It will be seen in subsequent chapters that, unless very special precautions are taken, it is a difficult matter to make the aeroplane fly over given ground in a given direction, in the presence of a cross wind; several of the runs to be studied bear evidence that the pilot, realising that he had not succeeded in doing this, had changed direction in the middle of the flight, in order to keep over the desired strip of country. In the surveys described in later chapters, great attention was paid to keeping the flights straight along their whole length, and it will be found there that the slow continuous curvatures, noticeable in the experiments now under discussion, have entirely disappeared.

The second figure (*B*) of each diagram shows the bearing, or azimuth orientation, of the centre line of the plate, which was itself parallel to the fore-and-aft axis of the aeroplane; each dot corresponds to the dot vertically above it in the plan figure, so that the horizontal spacing of the dots is of no importance, except to facilitate the location of the aeroplane in plan, at the moment in question. This last remark applies to all the lower figures; all the dots upon a vertical line on the paper correspond to the same exposure. The remarks as to gentle curvature of the track apply of course to this figure also; thus, the slow changes of azimuth that occur in these records can be avoided, if the proper precautions are taken. On the other hand, local fluctuations of azimuth, about a smooth curve drawn through the points, probably represent the type of fluctuation that may be expected, because the pilots were doing their best to avoid quick turns, on account of the lateral accelerations which they would involve.

The thin continuous line in figures *B* shows the bearing of the track over the ground, as measured from the successive plumb points. The probable error of its location is about one degree. The angle between the bearing of the aeroplane and the track made good over the ground is due, in the main, to cross winds, but it is possible for short period fluctuations to occur in the bearing of the aeroplane without being appreciably reflected in the bearing of the track.

Figures *C, D, E, F* contain the data for which the experiments were primarily carried out. In *C* is shown the lateral component of tilt of the camera, that is to say the angle made by the axis of the camera with a plane containing the vertical and the fore-and-aft axis of the aeroplane. It will be noticed that there are at least two dots to each exposure; these are the independent determinations of which mention has already been made. Since every determination is included in the figures, it is possible to see, at a glance, the relation between the accuracy with which the tilt can be determined and the magnitude of the tilts that actually occurred. The continuous line gives a rough indication of the tilt of the apparent vertical, as determined from the curvature of the track in figures *A* and the known speed of flight. This line is not at all accurately located, being liable to errors, the probable values of which vary between half a degree and one degree in the different flights. The object of including it in the figures is merely to indicate the order of magnitude of the tilt of the apparent vertical, due to the rather obvious curvatures of the tracks. It is possible by studying the relation of this curve to the actual recorded tilts to see whether the general curvature of the path, in any instance, has influenced the pilot's idea of the vertical. In most of the flights here recorded the slight curvatures of the path have apparently had no influence on the recorded tilt, from which it may be inferred that the pilots maintained their wings level laterally, more by reference to the horizon, than by their sense of balance, or their cross levels.

Figures *D* show the height at which each of the photographs was taken. As before, separate determinations of height from independent data on the same plates are shown separately. This figure is self-explanatory.

Figures *E* show the components of fore-and-aft tilt, or tilt about an axis perpendicular to the axis of the aeroplane and to the vertical. The continuous line, in this case, is the angle of climb or descent of the aeroplane, as determined from the successive heights shown in figures *D*. This line again is only located with a probable error of about one degree. Its interest lies in the fact that it enables one to see at once the extent to which the fore-and-aft tilt was due to the changing height.

Figures *F* show the magnitude of the actual tilt of the camera axis, without regard to the direction in which it occurred. The measured tilts are the components of the actual tilt and, since these components are at right angles to one another, the actual tilt is the root-mean-square of the measured components. The full dots show the tilts from the vertical and the small circles show the tilts from the mean direction of the camera axis during the flight in question. This mean is determined from the mean values of the component tilts. When the open circles show, on the whole, less tilt than the full dots, it is to be inferred that the initial adjustment of the camera axis was not sufficiently good for the purpose in hand, or that the pilot was flying with a permanent tilt.

Appendix II contains the pilots' reports concerning these flights, written immediately after returning from the flights and before the analysed results were available. These reports should be studied when examining the data in the diagrams; they give, amongst other things, an idea of the magnitude of the errors to which a pilot is sensitive. Let us now consider the individual flights and the accompanying pilots' reports.

Diagram 1 was the first successful flight in this series. In it the camera shutter failed to act on the thirteenth and fourteenth exposures, so that these are missing. Referring to figure *A*, it will be noticed that the flight consisted of two fairly straight runs with a definite change of direction near the seventh photograph; this may have been due to the pilot mistaking his instructions and following the wrong canal, but the precise reason for it is not known. Figure *B* shows a strong cross wind but is otherwise uninteresting.

Figure *C* shows that on the whole the aeroplane was flying with the right wing slightly higher than the left, and that this tilt increased progressively until it reached two degrees, when it was suddenly corrected. In the light of the evidence from later flights it is probable that this slow rise of the right wing was due to the aeroplane being out of trim laterally, so that the pilot would have to hold the right wing down by a muscular effort. When this occurs the muscles are liable to tire and the effort to be unconciously relaxed. One may suppose that the pilot, having his attention concentrated on other matters, did not notice this process occurring until the tilt had reached two degrees. In spite of this persistent bias in the direction of the tilt there is no tilt greater than two degrees.

Figure *D* shows a steady fall during the first half of the flight, which was noticed after the aeroplane had fallen 150 ft. It would seem that an adjustment was then made, either to the speed, or to the throttle, after which the aeroplane climbed until 50 ft. above the starting height, when a second adjustment appears to have been made. When considering this figure it must be remembered that the principal difficulty which the pilot experiences is that of attending to many different things at once; it would be possible to keep within closer limits of height if there were nothing else to distract his attention. This figure illustrates the importance of making very careful preliminary adjustments of the throttle, to ensure that the aeroplane, when flying at the required speed, is neither rising nor falling; it is probable that, in this case, the preliminary adjustment had not been carried out with sufficient care. Apart from the general fall and rise, due to inaccurate throttle adjustment, the minor fluctuations of height were very small indeed, being within the range of error of the calculations themselves.

In figure E, the fore-and-aft tilts show a slight tendency to follow the continuous line, which suggests that these tilts were influenced by the gradual fall and rise above noticed, but there are cases of erratic tilts not due to this cause. As a conjecture, it is possible that the relatively big tilt of the thirteenth photograph was due to the pilot's attention being engaged in altering his throttle, as suggested by the previous figure. Even this, the largest tilt, does not greatly exceed two degrees. It is noticeable that the mean fore-and-aft tilt in this flight is practically zero, showing that the preliminary adjustment of the camera had been very accurate.

Figure F shows the magnitude of the tilt of the camera axis from the vertical. Its main interest lies in the fact that only two of the tilts exceed two degrees. When the full dots, showing tilt from the vertical, are compared with the small circles, showing tilt from the mean position of the camera axis during the flight, it is clear that the slight bias noticed in the lateral tilt has not been sufficient to affect appreciably the order of accuracy of the results.

The pilot, in his report, makes no mention of any lack of lateral trim, such as is indicated by figure C, but he does record slight bumps, especially near the start of the run. He mentions that his method of flying straight was to use a sun shadow in the aeroplane.

The second and third analysed runs. Diagrams 2 and 3 were made by another pilot and were carried out consecutively in one flight. The results of this flight are distinctly better than the first; the pilot appears to have taken more care to get his aeroplane flying level and in the right direction before starting on the recorded run. Possibly the condition of the air was slightly more favourable, since figures B show no cross wind and his report states that the air was very steady. It is to be noted that this pilot used a different method of flying straight from that used in the first run; he flew upon a distant point instead of using a sun shadow. On the whole, as the result of further experience, we prefer the distant point method, although the shadow method is useful when the horizon is too hazy for the location of any definite point on which to fly.

In the first of these two runs the same tendency for the right wing to lift is just noticeable, but this pilot seems to have realised it sooner than the other, possible because the air was calmer. He appears to have slightly over-corrected on each occasion, so that the mean lateral tilt is in the direction of raising the left wing. The height has been kept within much narrower limits than before, the aeroplane having a tendency to loose height very slowly, but this tendency was noticed before it had caused a fall of more than 40 ft. and was corrected once in the middle of the run. The extreme variation of height is no more than

Diagram 1

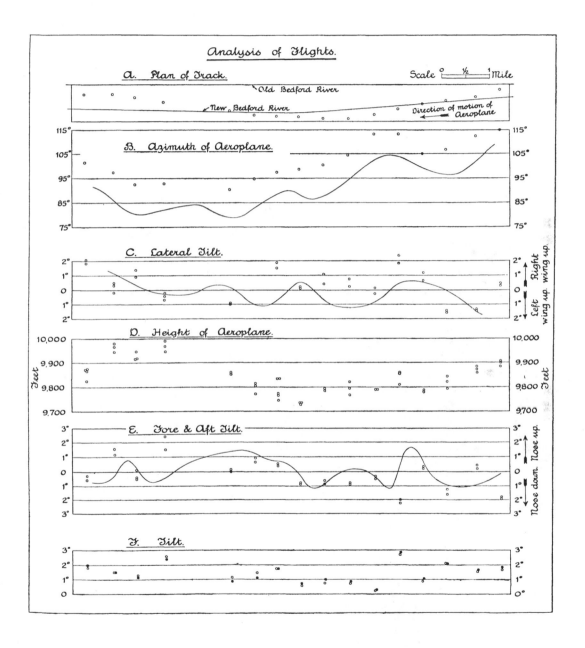

Analysis of Flights.

Diagram 2

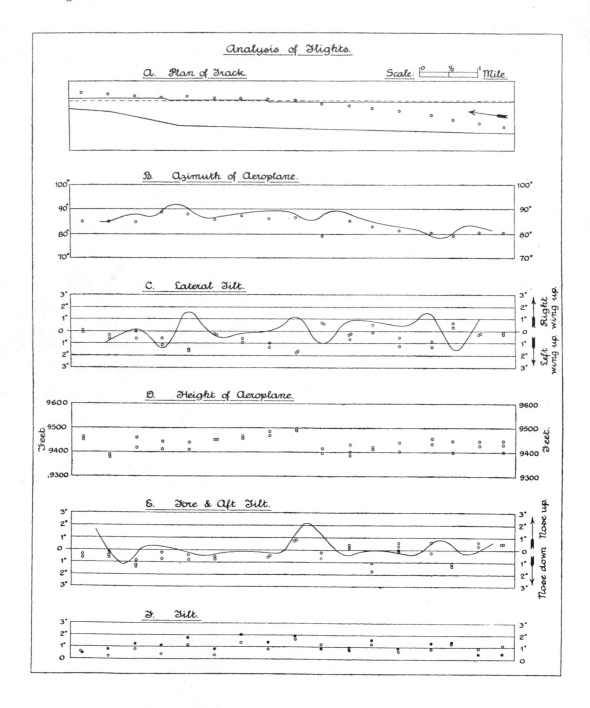

Diagram 3

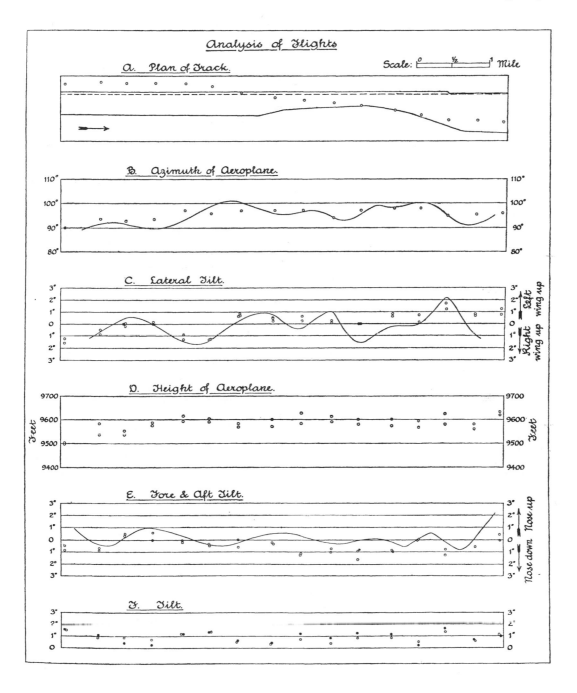

Diagram 4

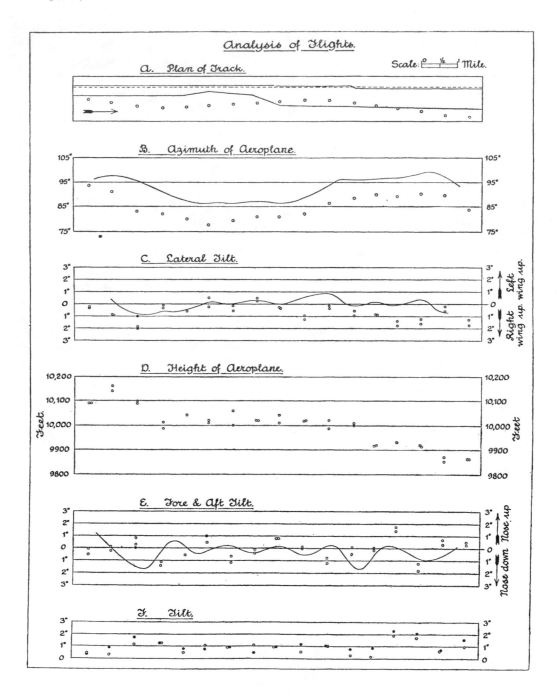

Diagram 5

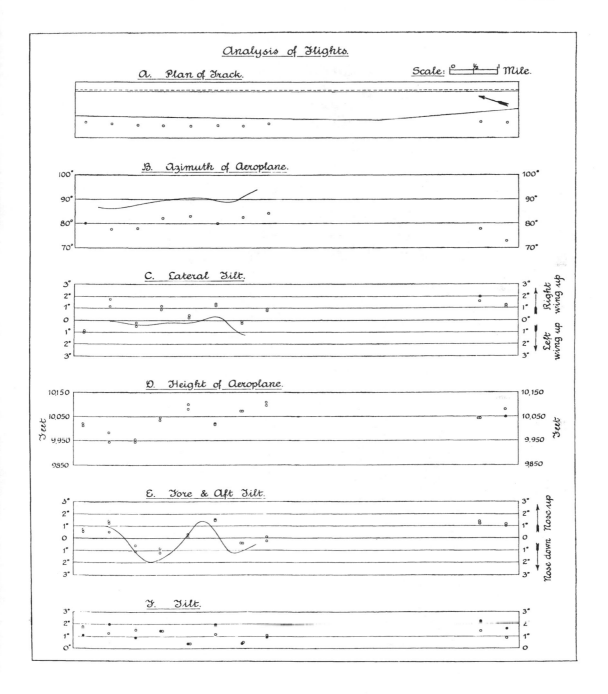

Diagram 6

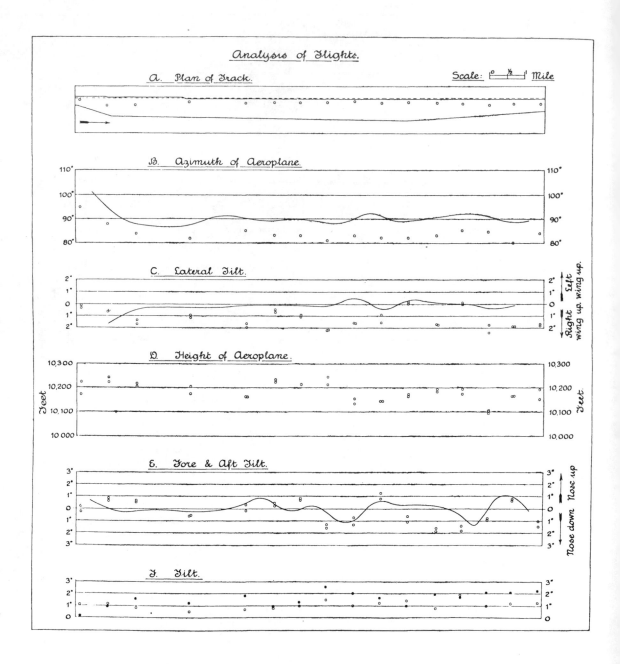

Diagram 7

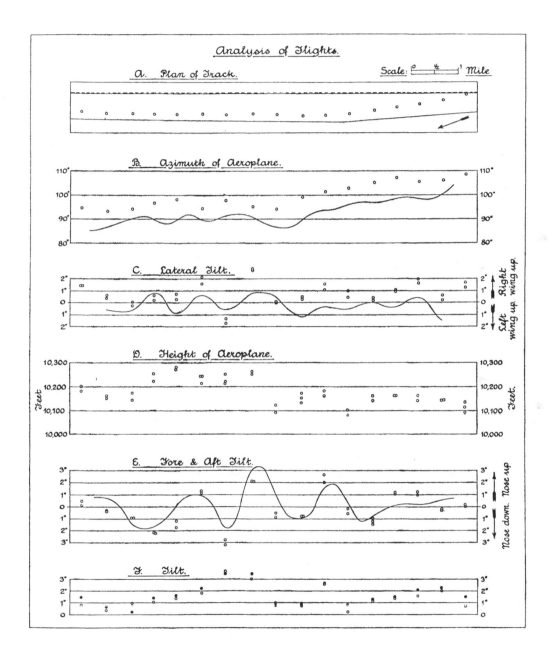

Diagram 8

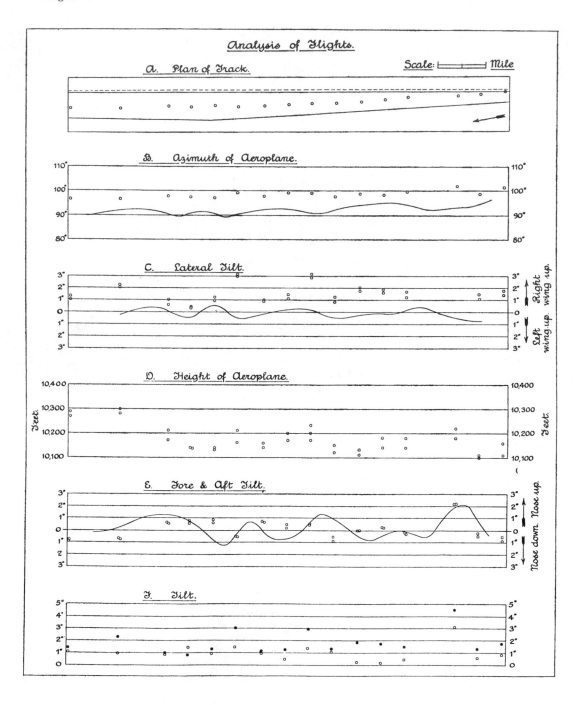

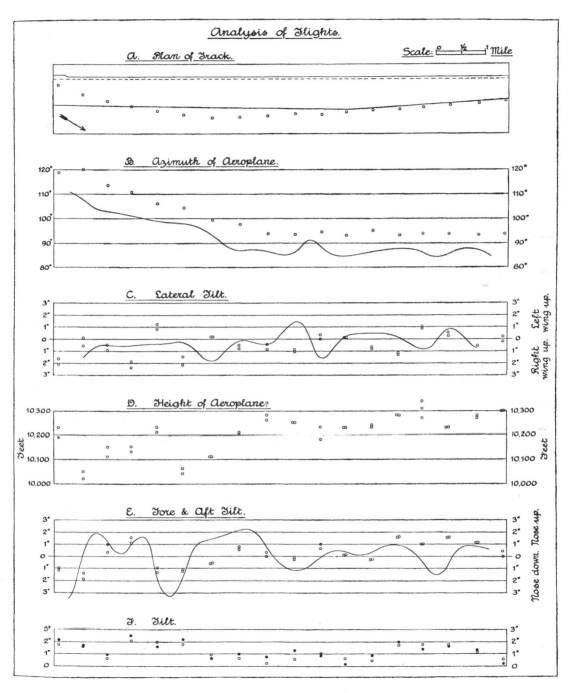

Diagram 9

Diagram 10

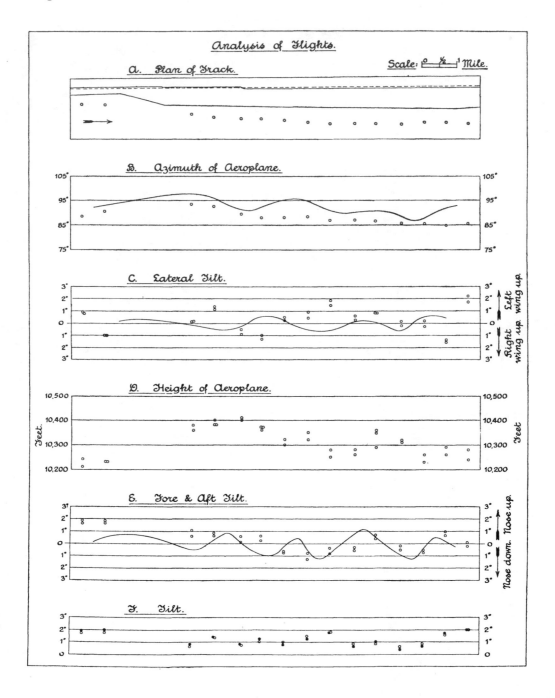

Diagram 11

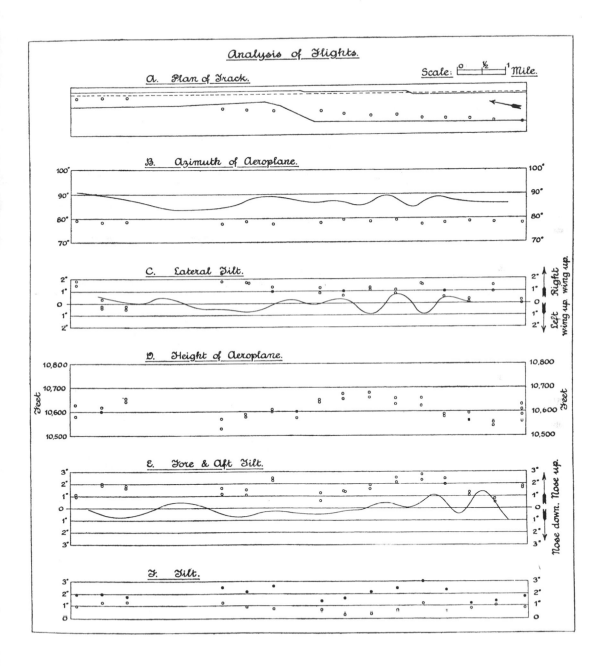

Diagram 12

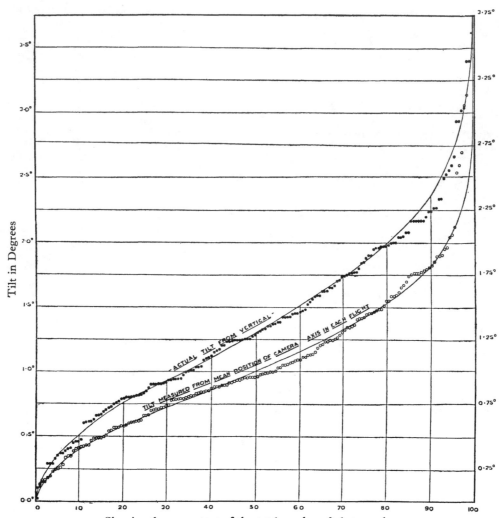

Showing the percentage of the total number of photographs
for which the tilt was less than any given angle.

Diagram 13

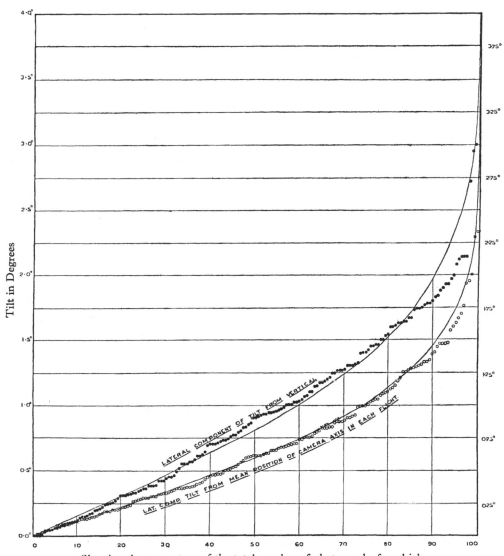

Showing the percentage of the total number of photographs for which
the lateral tilt was less than any given angle.

Diagram 14

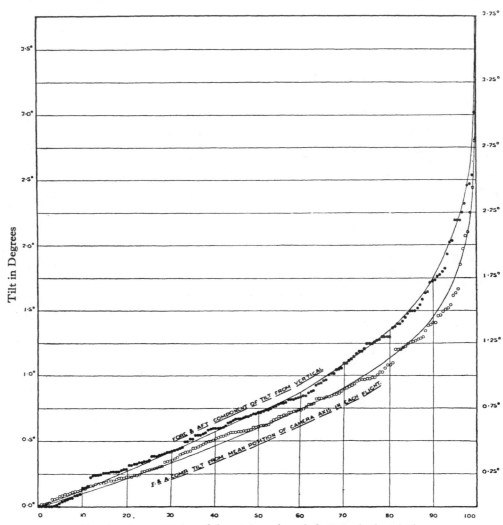

Tilt in Degrees

FORE & AFT COMPONENT OF TILT FROM VERTICAL

F. & A. COMP. TILT FROM MEAN POSITION OF CAMERA AXIS IN EACH FLIGHT.

Showing the percentage of the total number of photographs for which
the fore-and-aft tilt was less than any given angle.

Diagram 15

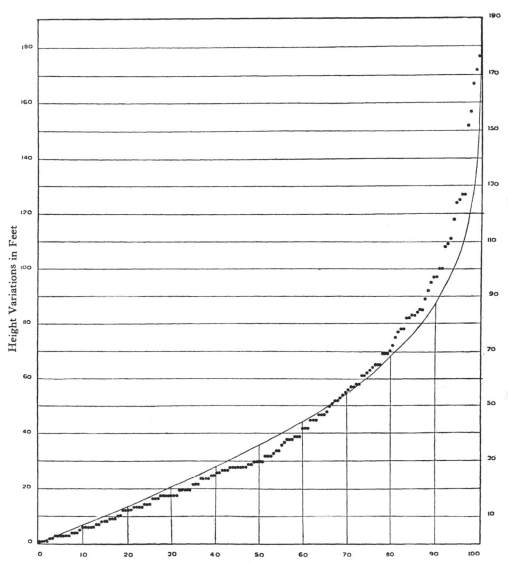

Showing the percentage of the total number of photographs for which the deviation of the height from the mean height of each flight was less than any given angle.

50 ft. on either side of the mean for the flight. The fore-and-aft tilts are very small and the total tilt never exceeds two degrees.

The second run of this flight, Diagram 3, is even better than the previous one. The height is amazingly constant, remaining well within 20 ft. of a mean value. Both components of tilt lie within one degree of the mean for each flight and there is no noticeable bias in any direction. The resultant tilts never reach two degrees and the average tilt is less than one degree. This is the best run of the whole series.

Diagrams 4, 5 and 6 contain three runs, made in a single flight by the pilot of Diagram 1. They are distinctly better than the first run, probably due to increased experience and to the fact that he had been able to study the results of his earlier attempt. The improvement in the flying is, however, masked by two factors. In the first place the tendency of the aeroplane to raise its right wing has now become sufficiently pronounced for the pilot to have noticed and recorded it in his report, and in all three runs there is a marked corresponding bias on the lateral tilts. Apart from this bias the lateral level is very well maintained, the fluctuations about the mean of each run never exceeding more than about one degree. It is worthy of remark that, had the camera been adjustable laterally, this bias would have been immediately noticed and corrected by an experienced observer, so that the actual tilts recorded would have been those from the mean. A glance at figures *F* will show that the results would thus have been considerably improved. In these three runs the camera had, for the first time, a screw adjustment on the fore-and-aft tilt, and it is noticeable that there is practically no bias on this component of tilt in any of the runs.

The other factor which has reduced the apparent accuracy of the results of this flight, is that the pilot experienced a difficulty, noted in his report, in adjusting his throttle. This has caused him to start his first run with the aeroplane falling, and his attention seems to have been so taken up with the other aspects of his work that he has not noticed this, even although he has fallen some 200 ft. This error is a good illustration of the necessity, to which reference will often be made, of relieving the pilot of the strain of continually attending to a number of different and delicate operations at the same time. When he has many things to attend to simultaneously, he is very liable to neglect one whilst attending to another which, for some reason, attracts his attention more strongly at the moment. Apart from these general tendencies to change height and to raise the right wing, these three runs are practically as good as the two in the previous flight. The large gap in the second of the three runs is due to troubles with the camera, which failed to act for a number of exposures.

Diagrams 7, 8 and 9 were the second, fourth and fifth runs of a single flight; the first has not been analysed and the third failed from camera troubles. The results are not quite so good as in the last two flights considered and occasional tilts up to three, and in one case four, degrees are recorded. The tendency to raise the right wing seems to be more pronounced than ever, but is still referred to by the pilot, in his report, as a "slight tendency." This pilot had great flying experience, and when he describes his aeroplane as being slightly out of trim it is probable that a less experienced pilot would notice nothing wrong at all. It would seem therefore to be of great importance to provide lateral adjustments on the camera itself, to enable the observer to compensate for consistent errors of this nature; they would be very noticeable upon a suitable spirit level carried on the camera. The height in these runs, especially Diagram 9, was more variable than usual; the cause of this does not appear in the pilot's report and is not known. The fore-and-aft tilt has also varied rather badly, apparently in sympathy with the changes of height. The resultant tilts are not good in these runs owing, mainly, to the heavy bias on the lateral components, but the variations from the mean position of the camera axis are of the same order as in previous flights.

Diagram 10 refers to the first of a series of three runs, the second and third of which have not been analysed. It has no special features of interest, but is a good run, of the order of accuracy determined in earlier experiments. Its value lies in confirming that these accuracies were not freaks which cannot be repeated.

Diagram 11 is not really comparable with the previous runs, since it was carried out with the help of a gyroscopic control on the rudder which automatically maintains the aeroplane on a straight course. In this flight the pilot's feet were off the rudder bar and his whole attention could be directed towards other operations, such as maintaining height and lateral level. In this run, there was a definite bias on both components of tilt, owing, probably, to the fact that the observer was new to the work and had not adjusted his camera correctly. If, however, the tilts from the mean direction of the camera axis are considered, it will be seen that the results of this flight for tilt, height and constancy of the bearing, are better than anything that has been previously recorded. The main value of this instrument is that it relieves the pilot of the strain of the most exacting operation of all, that of keeping continuously straight for long periods of time; it thus allows him, not only to concentrate more on other operations, such as keeping at a constant height, but to carry on longer in a single flight and so save time by reducing the total number of ascents required to survey a given area. It will be seen, therefore, that this instrument should be of great value in an aerial survey; it has, in fact, been extensively used in the experiments described in later chapters and further reference will there be made to it.

Now that we have extracted all the information we can from the flights considered separately, we shall proceed to consider the results in aggregate. For this purpose Diagrams 12 to 15 have been prepared. The method of studying groups of errors which is used throughout this book is discussed in Appendix I; where also is to be found a reference to these diagrams and the methods by which they have been constructed. It should suffice to state here that they show the percentage of the total number of photographs for which the tilt or height errors, as the case may be, did not exceed any assigned magnitude; and that, by comparing the chains of dots with the thin continuous lines, it is possible to see, at a glance, how nearly the distribution of the errors conforms to the "normal" distribution appropriate to errors due to entirely random causes. Each diagram refers to a set of observations of one kind, and the accuracy of a set is, for future reference, defined by the "50 per cent." error; that is to say the error which has a 50 per cent. chance of exceeding any individual error.

As in figures *F* of Diagrams 1 to 11 the dots in Diagram 12 represent tilts from the vertical, and the circles tilts from the mean direction of the camera axis in each flight. The fact that the latter are definitely smaller than the former gives a clear indication that the results might have been improved by the use of an adjustable camera. A re-examination of the data of the separate flights will make it clear that the difference between these two curves is almost entirely due to the tendency of the particular aeroplane used to raise its right wing. We are of the opinion that the chain of small circles will be found to represent closely the results that will be obtained with practiced pilots and adjustable cameras. For these the 50 per cent. error is less than one degree and the 90 per cent. error less than two degrees.

Diagrams 13 and 14 show the lateral and fore-and-aft components of tilt separately, the same convention being observed with regard to the dots and small circles. The "normal" distribution curves, added for comparative purposes, are of a different shape from those in the previous diagram, because there the tilt had freedom to occur in two directions, whereas here only one component of the tilt is recorded in each diagram. Two points of interest emerge from these diagrams. The difference between the tilt from the vertical and that from the mean direction of the camera axis in each flight is much greater in the case of lateral tilt than fore-and-aft tilt. This has already been attributed in part to the fact that, in most of the flights, the camera was adjustable in the fore-and-aft direction. The other point of interest is that the tilts from the mean direction of the camera axis are almost exactly the same in both directions. This is an important and somewhat surprising result, when the great difference between the control systems that are concerned in the two cases is taken into consideration. The result, however, does not stand alone, but is supported by almost every lengthy series of observations that have been made upon the oscillations of an aeroplane which is being flown as straight and steadily as possible. The 50 per cent. deviation from the mean is, for either of these components, 0·6 degree and the 90 per cent. deviation 1·4 degrees. The distribution is remarkably close to the "normal."

Diagram 15 shows the variations in height from the mean of each flight, the absolute variations from the supposed height were not measured. The distribution is not quite normal; the 50 per cent. error is 30 ft., and the 90 per cent. error 100 ft. It is probable that the slight preponderance of bad errors when compared with the normal distribution is due to the sticking throttle noticed in some of the reports. It is believed that better results than these could be obtained by more carefully adjusting the throttle so as to ensure, before the photography is started, that the aeroplane is neither gaining nor losing height. An inspection of Diagrams 1 to 11 suggests that, in many cases, this had not been done with sufficient accuracy.

Let us now consider what information of value to aerial surveying can be drawn from these experiments. In the first place, it has been shown definitely that it is possible to fly an aeroplane, of the type in question, in such a way that the tilt from the vertical of a camera carried in it seldom exceeds two degrees, whilst the height seldom varies more than 100 ft. from the mean value for the flight; the 50 per cent. value of these errors is less than one degree and 40 ft. respectively. This is a result of considerable importance, because it provides the fundamental data from which a preliminary estimate can be made of the accuracy to be expected from any type of aerial surveying, not employing re-section. It is important, therefore, to study the conditions under which these results were obtained and the probability of realising similar accuracies in the routine work of a survey.

In the first place, the flights were carried out on all occasions on which the aeroplane reached 10,000 ft. with a clear view of the ground, there having been no selection of the weather, except to ensure a clear view of the ground. It would seem, therefore, that it is probably a normal condition for the air, in clear weather at 10,000 ft., to be sufficiently steady for results of this accuracy to be obtainable. Subsequent experience has confirmed this conclusion, with the exception of one occasion when a strong north wind encountered at 10,000 ft. did not occur at low altitudes. On that day experiments of a somewhat different nature from those above had to be abandoned, because the wind was too high for the aeroplane to make any appreciable headway against it, and the pilot recorded that the air was too bumpy for him to have made a successful flight, even had he been able to make sufficient headway. This, however, is the only occasion, out of a large number of flights extending over three years and in all seasons, in which the air at 10,000 ft. has been seriously bumpy when a clear view of the ground is obtainable. It seems, therefore, to be fairly certain that, in England at 10,000 ft., the air conditions will generally be good enough to obtain the accuracies above recorded, whenever it is possible to carry out a survey at all.

All our work has been carried out at a height of 10,000 ft., that being a very convenient height for many reasons. We have little evidence as to the steadiness of the air at other heights, but we think that the air will not be more unsteady at heights greater than this, and that equally good work might be done from about 8,000 ft., or even lower. A descent to below 6,000 ft. might, however, introduce difficulties owing to "bumps," and we are sure that, below 4,000 ft. or below clouds, accuracies of the type we have recorded would be difficult to obtain with regularity; the air in these regions possessing, as a rule, quite a different order of turbulence from that usual at 10,000 ft.

The extent to which the turbulence of the air varies in different countries is not known to us and is a matter which might require consideration before a survey, depending on accurate flying, were undertaken. We think that, at 10,000 ft., one should be fairly clear of serious turbulence in clear weather, but that, if this is not the case, the trouble might be overcome by flying higher.

We have already mentioned that the accuracies obtained depend upon the skill of the pilot, and we think, from our experience, that a certain amount of special training will be required before pilots can obtain this order of accuracy. The difficulty lies, not so much in any one particular operation, as in the necessity of attending closely to a number of operations simultaneously, for an hour or so at a time, without seriously neglecting any one of them. We find that it is generally advisable to allow pilots to see and study the recorded results of their lapses of attention, and that after this they improve considerably. We have given reasons in an earlier part of this chapter why we consider that most good pilots will, after a little practice, be able to achieve an accuracy similar to that here recorded, and it will of course always be necessary to employ none but good pilots on commercial aerial surveys, simply from the point of view of safety to the machines and instruments.

Although we feel sure that any good pilot could be trained to produce results of this accuracy, we are also clear that the operations, when unaided by special instruments, are very exacting and tiring, and for this reason we are in favour of using any automatic apparatus that can be supplied for relieving the strain on the pilot, provided always that the apparatus can be quickly disengaged, should it fail, so that the pilot is enabled to carry on by his own skill and the flight is not wasted. This is a point to which we attach the greatest importance; we have already mentioned it at the end of the preceding chapter and shall refer to it again. The amount of work that can be carried out in a single flight and the cost of the flight are both so great that it is essential, in commercial work, to reduce the number of wasted flights to a minimum and, therefore, to keep the number of separate pieces of apparatus, upon which the success of the flight depends absolutely, as low as possible. It was in accordance with this idea that the preliminary experiments, with which this chapter deals, were nearly all carried out with the usual service equipment only. We consider that it will be sound practice, at least for some years to come, to train pilots to be able to carry out the necessary operations with the minimum of additional gear, but, in actual service, to equip them with all the aids that can be provided, such as gyroscopic rudder controls and automatic height maintainers.

Finally it is important to note that the results have been obtained under the easiest possible conditions, namely when the difficulty of finding the right track and keeping to it are reduced to a minimum; even so, the straightness of the tracks, although good enough for the purpose of the experiments, was not good enough for surveying. We are quite clear that a pilot who is attempting to ascertain whether he is passing over the correct ground during his flight, will not achieve results of an accuracy similar to that recorded. There still remains to be considered, therefore, the problems of covering the ground correctly and keeping the flights straight on the whole, by methods that do not occupy the pilot's attention during the time when the survey photographs are being taken. These problems will be discussed in the next chapters, where it will be found that it is possible to achieve flying of the accuracy of the present chapter throughout a real survey, but that in order to achieve this end great care is required in the choice of the methods by which it is ensured that the photograph will be taken from the desired positions.

CHAPTER V

MAPPING BY MOSAICS OF VERTICAL PHOTOGRAPHS

WE have seen, in chapter III, that in flat or gently undulating country it is possible to make a map out of vertical * overlapping photographs, without any precise knowledge of the position of the aeroplane, and without any ground control other than is necessary to prevent the accumulation of errors over great distances. This is done by taking all the photographs from the same height and fitting them together, to form what is called a "mosaic." A map made in this way may be surprisingly accurate in plan, but will not contain accurate height contours, although fairly effective form-lines can be sketched upon it by viewing overlapping photographs in a common stereoscope. In the present state of knowledge accurate height contours cannot be obtained from the air, except with the aid of elaborate instruments—such as the Auto-Cartograph—working from a close net of control points, which have been accurately located by ground surveying. There is, however, no reason why the same photographs which have been used in the first instance to make a mosaic map without accurate contours, should not be used subsequently in the Auto-Cartograph to give accurate contours when, in the process of time, it has become economically advisable to provide the necessary close and accurate ground survey.

The problem of making maps by the mosaic method can be separated into two well-defined parts: obtaining the photographs and combining them into a map. Let us deal first with the obtaining of the photographs. The aeroplane must be flown so that the ground becomes entirely covered by photographs which overlap each other and are all taken, as nearly as possible, from the same height and with the axis of the camera vertical. We saw in the preceding chapter that it is possible to fly consistently so as to keep the changes of height and the tilt of the camera axis within certain limits, and it is not difficult to show that, if this accuracy of flying could be realised in the actual mapping flights, it would be sufficient for the production of very useful maps. We saw, however, that this order of accuracy can only be obtained when the pilot gives all his attention to flying, and is not required to look at the ground, or to alter course during the flight.

It is essential, therefore, to devise a scheme of work which will not compel the pilot to change course during a photographic flight, and this,

* Photographs taken with the axis of the camera approximately vertical.

as was pointed out in chapter II, implies a knowledge of the compass course that will carry him in the required direction, which again implies a knowledge of the wind at the height of the survey. The track, moreover, must be located in position as well as direction, and this introduces all the difficulties, enumerated in chapter II, of locating, and flying over, a pre-arranged point.

The details of the flying routines and of the instruments used in these operations are fully described in chapter VII, and it is therefore proposed to describe here the general outline only of the routine to which we have finally come, and the influence that it has upon the course taken by the survey and the area covered in one flight.

The first problem to be surmounted is that of identifying the starting-points of the consecutive strips of photographs, which we shall, in future, describe as the "mosaic strips." In previously mapped country this presents no serious difficulty, but it is a problem which requires very careful consideration in undeveloped and unmapped country. In such country we should begin the survey by constructing a photographic strip at right angles to the direction in which it is intended to make the mosaic strips, and situated so as to contain the starting-points of these strips. This preliminary strip would be developed, mounted, marked with the starting-points, and given to the pilot to be taken in the air on his main surveying flight. It would of course be very wasteful if the pilot took photographs for his mosaic when flying in one direction only; so that it is advisable to give him two such "indication" strips, one to show the starting-points for the out flights and the other for the return flights. It is therefore necessary to decide, in the first place, how far these indication strips shall be spaced apart, or in other words, how long the individual mosaic strips are to be.

The choice of length for the mosaic strips is governed by the fact that they are located only at one end, and that their direction is determined with a finite accuracy only. If they are too long, gaps are bound to appear between them, and the length that can be given to them, without too great a danger of leaving gaps, is a complicated function of the lateral overlap allowed, the accuracy of location of the starting-points, and the accuracy of the direction of flight over the ground. In our experimental surveys we have been able, with a 30 per cent. overlap, to use mosaic strips fifteen miles long without leaving gaps, but, in view of the extreme importance of not leaving gaps in a commercial survey, we consider ten miles to be the ideal length for mosaic strips. This length has the additional advantage that, with it, a convenient day's work of about one hundred square miles forms a square of ten miles to the side which provides a very compact and convenient unit for compilation.

The preliminary indication strips will thus be spaced about ten miles apart. Since it would not be economical to spend a whole flight in making the two ten mile indication strips required for one day's mosaic mapping, the obvious procedure would be to make the indication strips for several days' mapping in one flight. We consider therefore that the first step in constructing a large aerial map by these methods would be to divide the country up by a series of long identification strips, spaced about ten miles apart, and each about fifty miles long. Two such strips could be easily made in one flight.

The methods used in laying down these preliminary strips, and the extent to which they might be used to give some control over the position of the separate mosaics, when the ground survey control points are very widely spaced, is described in chapter VI; there also is to be found a suggestion for the way in which widely spaced flights of this nature could be used, by themselves, to provide a reconnaissance map of large areas of country at very low cost.

In some types of country, where there is very little detail, it may not be easy to locate the starting-points for the mosaic strips upon the identification strips, but it is difficult to see what better indication can be given to a pilot than an actual photograph of the spot to be located; it would seem that, when there is not sufficient detail on the ground for the pilot to locate his starting-points by means of photographs, the country is not suitable for mosaic mapping. Such country, in fact, would hardly repay detail mapping of any kind, and in such cases it is probable that there would be no incentive to carry an aerial survey further than the reconnaissance stage discussed in chapter VI.

The procedure for photographing the mosaic strips, which we finally adopted after considerable experiment, is as follows. The first operation on reaching the survey height, which with us was 10,000 ft., is to trim the controls and adjust the throttle until the aeroplane flies at a constant height, at the desired speed, with a minimum of attention. We next find the direction and strength of the wind, and, from this, calculate the compass course that must be flown, in order to make the mosaic strips lie parallel and in the desired direction, on the out and return journeys. The same data is also used to calculate the times between exposures which will give the desired fore-and-aft overlap on both journeys, and the angles through which the camera must be rotated in azimuth to cause the edges of the photographs to lie perpendicular and parallel to the direction of the strips. This last adjustment is more important than might be supposed; if it is neglected, the effective width of the strip may be much reduced when there is a strong cross wind. The illustration facing p. 55, which shows two strips, in one of which the adjustment for drift angle has been made whilst in the other it has not, should make this point clear. The bottom strip in this illustration is an example of an identification strip marked ready for the air.

These operations are described in detail in chapter VII, but we may say here that, in our experiments, the wind was found by the method of flying on three courses, making 120 degrees with each other, and noting the drift of the aeroplane in each case. The instrument used for this purpose was the wind gauge bearing plate, which is used in the Royal Air Force in connection with the navigation of aeroplanes. With a practised crew, the total time required for this operation lies between ten and fifteen minutes.

When the necessary data concerning the wind has been obtained, the pilot approaches the starting-point of his first strip, and endeavours to get his aeroplane vertically over it, whilst flying by compass on the course communicated to him by the observer. He is instructed to get over the point at all costs, even if it means deviating slightly from the desired course, and to return to his course as quickly as possible after getting over the point. The fundamental difficulties of this operation are discussed thoroughly in chapter II, and the practical aspects of the problem in chapter VII.

The pilot signals to the observer as soon as he has arrived over the starting-point, whereupon the observer makes his first exposure and, henceforward, exposes at regular time intervals, which depend on his previous calculations. During the flight, the pilot concentrates upon keeping the speed and height constant, the aeroplane on a straight course, and the lateral level of the wings correct; he never looks at the ground beneath him. The run is ended at a signal from the observer that the pre-arranged number of exposures has been made, and the pilot and observer are now at liberty to look down and locate themselves with reference to their position on the second indication strip; if they have not arrived over the exact spot which is marked as the end of their mosaic strip, they can make notes as to the corrections to be made to the course and the exposing interval on the next run in the same direction.

After turning the pilot picks up the start of the return strip, and the process is repeated with, in general, a different time interval between exposures and a different drift angle on the camera, owing to the different effects of the wind when flying in opposite directions. The whole process is continued, without a break, until the day's work is completed.

If the automatic gyro rudder control, of which mention has been made, is used, it is switched into action after each strip has been well started and the aeroplane has settled down on the required course; if the apparatus is working well, the pilot need not give any further attention to the direction in which he is flying, beyond an occasional glance at his compass to verify that the gyroscope is not precessing. He is therefore free to concentrate more closely upon maintaining height, speed, and lateral level correct.

We have explained in a previous chapter that all existing compasses are seriously influenced when the aeroplane turns, and are not to be relied upon until the aeroplane has been flying in a straight line for from one to two minutes. This defect introduces great difficulties into mosaic surveying, because it makes it necessary to waste valuable time in flying straight for one or two minutes, before a flight by compass can be relied upon to be in the desired direction. Even after this preliminary straight flight has been made, it is often necessary to make some slight turn on approaching the starting-point, in order to get over it accurately; this turn will disturb the compass and hence cause the first part of the mosaic strip to be in the wrong direction; this again will cause the latter parts of the strip to be displaced laterally, and thus increase the liability to leave gaps between successive strips.

To counterbalance this defect in the compass we use a second gyroscopical instrument known as the "gyro azimuth indicator." This consists of a balanced gyro similar to that used in the gyro rudder control, but in this case there are no relays, the gyro being directly coupled to a pointer which traverses a scale of degrees in a similar manner to the needle of a compass. This gyro and pointer can be locked by the pilot so that the pointer indicates zero. To use the instrument the gyro is unlocked when the aeroplane is flying level and on the required bearing by compass; whereupon the pointer will, henceforward, indicate zero whenever the aeroplane is flying on this original bearing, and at other times will indicate variations from it. This instrument has, of course, no absolute sense of direction and will go wrong after a time, owing to slight precessions in the gyro, but the specimen which we have used maintains its original orientation with sufficient accuracy for over ten minutes, and during this time is of immense value on account of the fact that it is quite undisturbed when the aeroplane turns.

We have found, in practice, that these two gyroscopical instruments are of such great value, both from the point of view of saving time and of increasing accuracy, that we should recommend their use in any large scheme of aerial surveying, in spite of the fact that it is possible to obtain moderately good work without them. Both these instruments were designed and made by the Royal Aircraft Establishment and lent to us for this work.

The above is but a brief outline of the flying operations necessary for the construction of a large mosaic map. In practice the routine to be followed in the air must be worked out to the minutest detail, the pilot and observer being provided with definite instructions and blank forms to be filled in, which leave no necessity for thinking in the air. The details of these routine instructions and of the forms and curves to facilitate calculation, together with more detailed descriptions of instruments employed and of the precise methods by which they are used, will be found in chapter VII.

Having considered briefly the processes by which the photographs required for a mosaic map can be obtained, we shall now turn our attention to the compilation of the mosaic in the office. Our object is to devise methods which will be cheap and rapid enough for use in the construction of maps of large areas of country, at a very low cost per square mile. We aim, therefore, at using methods in which the contact print, direct from the original negative, can be used, avoiding methods that require the re-projection of individual photographs at different angles and to different scales. We assume that the flying has been done according to the methods which we have indicated and that the photographs have, therefore, all been taken from approximately the same height and with the axis of the camera very nearly vertical. We also assume that local variations in the height of the ground do not exceed a few hundred feet.

If the ground were absolutely flat and the flying absolutely accurate, it would be possible to fit all the photographs together to form an accurate map with no discontinuities of detail between the photographs; but, in practice, changes of height in the ground, and slight inaccuracies of the flying, will introduce small differences of scale between different photographs and between different parts of the same photograph, which will make exact fitting impossible. When only two or three photographs are to be fitted together, these differences should be scarcely noticeable, but, when a large mosaic is being compiled, the discrepancies tend to accumulate and produce two undesirable results; in the first place regions occur where, owing to the accumulation of errors, it is impossible to get even an approximate fit between two consecutive photographs, and in the second place the general form of the map becomes distorted. If the mosaic to be made is a small one, involving not more than ten or twenty photographs, it is often possible to distribute the small errors satisfactorily by pure skill and experience, but where a large mosaic of one hundred or more photographs is to be made, we have found that it is absolutely necessary to adopt systematic and more or less mechanical methods for the distribution of errors. We shall proceed to describe the method which, as the result of considerable experience, we have finally adopted.

Let us assume that we desire to construct a mosaic of about 100 square miles, in the form of a square of 10 miles to the side. The photographs which we are using are 5 by 4 inches in the sides, with the five-inch length across the direction of the strips. They are taken with a focal length of 6 inches, from a height of 10,000 ft., and therefore cover a rectangle on the ground of 1·6 by 1·3 miles. We have aimed at a 30 per cent. overlap in all directions, hence, making allowance for contingencies, there should be about 13 photographs to a mosaic strip, and about 10 strips to the 100 square miles. We shall have, therefore, to deal with some 130 photographs taken in strips of 13. These photographs will represent the ground on a scale of about 1/20,000 or roughly 3 inches to the mile, so that our finished mosaic will be about 30 inches square. It is probable that large surveying schemes to be attempted in the future will be carried out with film cameras, giving photographs 10 by 8 inches to the side, but having about the same angular cover as the camera used by us. With these photographs the numbers given above would remain unaltered, with the exception that the mosaic would be 60 instead of 30 inches to the side.

In compiling the mosaic we make use of the fact that the separate strips will have been taken in very straight lines, and we begin by laying out each of these strips separately upon a table. Starting with the first photograph of the strip, we fit the next to it by means of the detail at the edges, and proceed in this way until the whole strip is laid down. If the strip has been taken in an exact straight line, and if the axis of the camera has been vertical for each photograph and the ground flat, this process will result in a continuous band of photographic prints, the centres of which lie in a very straight line; but, if all these conditions have not been fulfilled, the strip of prints so formed may show a pronounced curvature. This curvature may be due either to the fact that the centres of the photographs actually lie on a curve on the ground, or to the camera having had a consistent lateral tilt relative to the ground, which caused the prints to represent the ground to a larger scale on one side than the other. We describe curvature due to the first cause as "real" and that due to the second as "fictitious." In practice the curvature will be partially real and partially fictitious, and it is part of the process of accurate compilation to eliminate the fictitious curvature by distributing errors between the joins of the individual photographs, whilst preserving the real curvature which should, of course, appear in the final mosaic.

The illustration facing p. 55 shows two such strips; one of these is nearly straight, except near one end, and one has a pronounced curvature. In both cases the curvature is nearly all fictitious and is due to the presence of a slope in the ground. When in their final position in the mosaic, these two strips are very straight.

Again, the distance between two given points on the strips of prints obviously depends upon the average height of the camera above the ground during the flight on which the photographs were taken. If this average height varies between adjacent strips, the distance between any two points that appear in the overlapping parts of both strips will be represented by different lengths in each strip. In a ten mile strip, for instance, a discrepancy of this nature amounting to about 0·05 mile might occur when the average height in the two strips differed by only 50 ft. We have seen that errors of over 50 ft. may occur in the height of the aeroplane, hence some process is required for adjusting the total length of individual strips before they can be made to fit together laterally. This adjustment could be made by passing the individual prints through an enlarging lantern arranged to give the required alteration in scale, but this would be costly in time and material, and it is generally sufficient merely to distribute the errors so that there is no excessive discrepancy between any two prints. The case mentioned would be satisfactorily dealt with by a reduction in length of 0·025 mile on one strip and an increase of 0·025 mile on the other; this could be effected by adjustments of about 0·002 mile each between individual prints.

Before the separately compiled strips of prints can be successfully fitted together, we are thus faced with the problem of eliminating both their fictitious curvatures and their errors of scale, by means of small adjustments between successive prints. When it is remembered that we cannot distinguish fictitious from real curvature, except by the process of trying adjacent strips together, and that we do not know which of the strips, if any, is to the required scale, it will be seen that the adjustment is not one which can be easily carried out by entirely haphazard methods. We have tried several ways of attacking this problem of compilation on a large scale, and have finally adopted the following, as a routine process.

Having, as already described, laid out the strips of prints separately, we pin them down along one edge and slip beneath them two stretched elastic bands, the ends of which can be seen protruding from the curved strip in the illustration facing p. 54. Each of these elastic bands is secured to each print, by a small dab of seccotine, and the print weighted down until the seccotine is set. The whole arrangement is then released from the table and becomes a chain of prints, connected together by the two elastic bands only.

When all the strips that go to a single mosaic have been treated in this way they are removed to the compiling table (the illustration facing p. 54) which merely consists of a flat table carrying the material on which

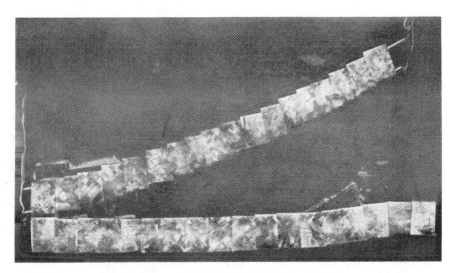

Showing photographic strips with fictitious curvature

Showing process of compilation of mosaic without independent control

The mosaic on the compiling table showing process of
compiling mosaic with independent control

Showing, above an "indication strip," below the effect of omitting
to rotate the camera to allow for "drift"

the mosaic is to be compiled. The chains of photographs are laid on this table in order, roughly as they will appear in the mosaic, and the ends of the elastic bands pinned in place.

The next step is to choose a good strip near the middle of the mosaic, and stretch it until the detail at the edges of the prints runs continuously over the joins, as when it was first laid out. The adjacent strips on both sides are then stretched in a similar manner, but it will generally be found that, when this has been done, they will not fit the first strip, because of their differing fictitious curvatures and lengths. Small adjustments are then made to the strips by lengthening and shortening the elastic bands, and by dragging the middles of the strips sideways and securing them by means of lead weights; no attempt is made, at this stage, to adjust individual prints. More and more strips are then adjusted in this way, until the edges of the mosaic are reached. As each new strip is brought into the adjustment, it provides an additional check upon the curvatures of the ones before, so that it is continually necessary to revise slightly the adjustments that have already been made; this does not take very long, provided that the adjustments at this stage are confined to the total length of the strips, and to the lateral position of a few points only on each strip. When this preliminary general adjustment is properly carried out, it results in the almost entire elimination of fictitious curvatures, and it adjusts the relative positions of widely separated prints with surprising accuracy. It is very important that the general adjustment be thoroughly carried out before any detail adjustment between individual prints is attempted, otherwise an immense amount of time may be wasted.

When the general adjustment has been completed, all the prints of the mosaic will be approximately in their correct relative positions, but there will be discontinuities between individual prints, larger than need be accepted, and which somewhat spoil the appearance of the whole. A detail adjustment can now be carried out, from print to print, taking great care not to move the prints far from the positions they have taken up in the general adjustment. During this stage small errors may be found in the general adjustment, which, when reduced, may allow a better fit to be obtained, and when this is the case, a slight general re-adjustment should be made, but great care should be taken not to do this until quite certain that it is necessary, and that the change will not introduce worse errors in some other part of the mosaic.

A mosaic at this stage of the compilation is illustrated facing p. 54. The elastic bands are here clearly seen. The prints in this illustration have not yet been trimmed and are only lightly secured upon the board.

When the detail adjustment has been made, the mosaic will have reached the highest stage of accuracy that can be obtained without the use of control points. The scale of the whole, will, of course, depend on the average height of the camera above the ground during the photography, and the distortion will depend upon several factors such as the accuracy of the flying, the levelness of the ground and the thoroughness of the compilation. We have carried out an analysis of two mosaics after they had reached this stage, and before any adjustment had been made to known control points. These mosaics had each an area of over 100 square miles, but instead of being square, as suggested, formed rectangles of about $7\frac{1}{2}$ by 15 miles to the sides, and each consisted of about seven strips of 15 miles in length.

In order to study the distortion of these mosaics, we selected some 40 points on each, which could be identified on the mosaic and on the ordnance, one inch to the mile, coloured maps. The positions of these points on the mosaic were plotted on transparent paper, to a suitable scale, and a corresponding plotting made from the map, to a scale that gave the best possible fit between the two groups of points. Had the mosaic and the map had no distortion whatever, it would have been possible to obtain an exact fit between the two sets of points when one was laid over the other. In practice, owing to distortion, an exact fit could not be obtained, and the discrepancies outstanding, after the best possible fit had been obtained, give a measure of the relative distortion of the map and the mosaic; or, assuming that the map is not distorted at all, a measure of the distortion of the mosaic. The ordnance maps which we used were, as a matter of fact, considerably distorted, owing to shrinkage of the material on which they are mounted, but this shrinkage is fairly uniform in any given direction, so that it was possible to allow for it; we think it is safe to assume that the positions of the points determined from the map represent the actual configuration of the ground with sufficient accuracy for the purpose in hand.

As the result of these comparisons it was found that, in the first mosaic, which was carried out without the help of the gyro rudder control but with the gyro azimuth indicator, most of the errors were less than 0·05 mile but that, towards the ends of the mosaic, there were more or less systematic displacements amounting in some cases to 0·15 mile. In the second mosaic, which was the same size as the first, but in which both gyro azimuth indicator and gyro rudder control were used, no point was displaced from its true position by more than 0·05 mile. The improved accuracy in this case was mainly due to the increased accuracy of the flying, resulting from the use of the gyro rudder control. The flights were straighter, and the tilts, as evidenced by the fitting of adjacent photographs, were less; the compilation was therefore considerably easier than in the mosaic in which the rudder control was not used. It is this mosaic that is shown in the illustration facing p. 54.

The process of comparison just described does not show up errors in the average scale on which the mosaic was constructed, but it was of course necessary to find this scale before the comparison of distortions could be made. This was done by comparing the distances between a number of widely spaced points on the map and mosaic. The intended scale of the mosaics was 1/20,000, and the actual average scale, in both cases, was, as near as could be measured, 1/19,800, or 1 per cent. too large. This may have been due to the aeroplane having been about 100 ft. too low, but it might be due, in part or in whole, to shrinkage in the prints, or to systematic errors introduced during the compilation. It was stated in chapter III that height, determined by aneroid, with an adjustment for temperature at the height, should be correct within 1 per cent. if care is taken, and all that can be said from these two results is that they lie within this error. The flying for the mosaics was carried out on two successive days, so that the temperature gradients and the errors introduced thereby might be expected to be approximately the same in both cases.

So much for the results that can be reached without employing any ground control whatever. In most practical work, however, some ground control will be used, either from existing triangulations, or from a special ground survey. The method of using this ground control will depend upon the spacing of the control points; if these are so closely spaced that two or more occur in each mosaic, they can be used directly, during the compilation of the mosaic, to check its scale, its orientation and to some extent its distortion. If, on the other hand, they are much more widely spaced than this, they can only be used in conjunction with some larger scheme, such as that described in chapter VI. We shall consider now the case in which the ground control is sufficiently close for several control points to appear on a single mosaic. We have already given it as our opinion that it is not advisable to attempt the construction of mosaics of more than 100 square miles area, so that we shall now be considering the procedure suitable to a ground control in which the control points are not more than ten miles apart.

The most economical and satisfactory ground control, from the compilation point of view, would be one in which the control points form a square grid, ten miles to the mesh, and situated so that the points fall on the corners of the mosaics; for in this case each point would directly control four mosaics, and each mosaic would be controlled by four points; and this end would have been attained with the minimum number of control points, namely, one per hundred square miles. In practice it may seldom be possible to provide a control of this nature, owing to the fact that triangulation points cannot, in general, be chosen at will, but must be restricted to points from which a good view can be obtained.

When we wish to compile a mosaic to control points, we start as already described, but instead of fastening the ends of the elastic bands directly to the table, as in the illustration facing p. 54, we fasten them to two wooden laths clamped to the edges of the table, as in the illustration facing p. 55. These laths are supported just clear of the table by thin packing pieces, so that the elastic bands can be passed beneath them, turned over and pinned, as illustrated. The laths can be unclamped and moved bodily, carrying the ends of all the elastic bands with them, and in this way the whole mosaic can be subjected to systematic linear strains, at any stage of the compilation. The mosaic strips are set up between these laths and given the first general adjustment which determines the shape of the mosaic, but, before proceeding to the detail adjustment, the average scale of the mosaic is found by comparing distances between control points on the ground with corresponding distances in the mosaic and averaging the results of these comparisons.

A template of the control points, on this average scale, is next constructed, consisting of a light wooden frame carrying small squares of transparent celluloid, upon which the positions of the control points are marked. This template is laid on the mosaic and moved about until the best possible fit is obtained. Owing to distortion the fit will not be exact and the remaining discrepancies must be removed by systematic straining. Strains in the direction of the strips are very easily applied, by moving the laths to which they are fastened, but strains at right angles to the direction of the strips are not so easy to apply in a systematic manner. In our own experiments we adopted the plan of laying down a series of parallel elastic bands—seen as black bands in the illustration—beneath and at right angles to the bands on which the photographs were fastened. At the end of the first general adjustment these transverse bands were secured in various places to the bands carrying the prints, the idea being that, when the former were stretched, the mosaic would experience systematic strains at right angles to the photographic strips, and that these strains would be distributed proportionally over the whole width of the mosaic.

These transverse bands did not work very well, possibly because they were too strong and not sufficiently elastic. The longitudinal bands on the other hand were very satisfactory and of about the correct stiffness for the purpose; in future work the transverse bands might be made of the same material as the longitudinals. In our experiments the longitudinals were made of an article of trade known as "hat elastic" whereas the transverse bands were made of "garter elastic." We considered that there was scope for considerable improvement in the method of distributing transverse strains; our method was not altogether satisfactory, although the fact that it was workable is proved by the accuracy of the finished mosaic which accompanies this book.

The systematic stretching of the mosaic should result in the control points on the mosaic coinciding exactly with the corresponding points on the template, and thus being correctly placed in relation to each other. The prints containing these control points should therefore be pinned finally in place, to ensure that they do not move during the subsequent adjustments. If there are not more than four control points, they can all be dragged into place merely by moving the four laths to which the ends of the longitudinal and transverse elastic bands are attached, but, if the control points number more than four, some of them may have to be pulled into place by an application of force to the prints containing them. In the controlled mosaic that we compiled, we only assumed four control points, so that the whole of the stretching was carried out by means of the laths, without actually touching any of the prints. If the mosaic has been carefully constructed the actual amount of the straining need only be very small. In our mosaic the biggest strain required amounted to about 1 per cent.; the control points were about 10 miles apart on the ground and about 30 inches on the mosaic, so that the greatest distance that any control point had to be moved, from the position taken up in the first uncontrolled compilation, was of the order of 0·15 inch.

When the controlled points have been fixed in their correct relative positions, the mosaic must be given a final adjustment; first the general adjustment must be checked over, to see whether any errors have crept in during the straining, and finally a detail adjustment of individual prints must be made, taking care as before not to disturb the general adjustment, unless absolutely necessary. As each area receives its final detail adjustment the prints should be held firmly in place by weights, so that, when the adjustment is complete, the whole mosaic will be covered with weights, as in the illustration. The lead weights which we used to hold the prints in place were specially made for the purpose; they were rectangular in form, about three quarters of an inch square by an inch and a quarter long. When standing on end on the prints, these weights have considerable holding power and do not cover too much of the surface of the print; a large number of them can be seen in the illustration. The boxes of photographic plates, also seen in the illustration, were used to replace the weights in areas which had been completely finished. When equipping an office for continuous compilation by these methods, it would be advisable to provide, for each mosaic, several hundred of the small weights and fifty or more much larger weights to take the place of the boxes of plates.

During the compilation of the mosaic care must be taken that the prints overlap each other according to an ordered plan; thus successive prints in each strip, and successive strips must all overlap each other in the same directions. If this has been done, it will be possible, starting from that corner print which is not overlapped by any other, to pick up all the prints in succession, without disturbing any of those that remain. It is, of course, necessary to remove the elastic bands from beneath the prints before finally trimming and pasting down, and this can be done in the following manner. Starting with that corner print which is not overlapped by any other, mark, on the base board, the position of the edges that do not overlie other prints; disconnect this print from the elastic bands to which it is fastened and remove it from the board, after marking it with an identification number to correspond with a number written on the board itself. Proceeding in this way the whole board can be cleared of prints, and the elastic bands removed. The prints can then be trimmed, replaced in the reverse order and pasted in place, care being taken to avoid trimming away those edges of the prints that have been used to fix their position on the board. The temptation to attempt any further adjustment of position during this stage must be firmly resisted; the prints should be replaced exactly in the positions from which they were removed.

The mosaic is now finished, so far as the compilation is concerned, but, in all probability, it will not be on exactly the desired scale to which the whole map of many mosaics is to be constructed. The subsequent procedure will depend upon the type of map that is required, but it will probably be desirable, in most cases, to re-photograph the whole mosaic in a large copying camera and, in doing so, to adjust it to the required scale.

In order to gain experience on this problem of compiling to control points, we provided ourselves with a new set of prints of the two mosaics to which reference has already been made; and, starting again right from the beginning, compiled them into a single controlled mosaic of 15 miles square, or 225 square miles. The resulting controlled mosaic has been reproduced on about half the scale of the original prints, and is carried in a pocket in the cover of this book, whilst a graphical representation of the errors to be found in the original is shown on a separate sheet. In this the centres of the four double circles represent the control points to which the mosaic was stretched, whilst the vectors inside the single circles represent the displacement of points in the mosaic relative to corresponding points on the one inch to the mile coloured ordnance map. The centres of the circles are the positions of the comparison points as taken from the ordnance map, and the ends of the vectors are their positions as taken from the mosaic. The magnitude of the errors can be easily estimated by comparison with the radius of the circle surrounding it, which is one-tenth of a mile.

The Northern half of the mosaic, containing the town of Cambridge, was carried out without the help of the gyro rudder control, whereas in the Southern half, containing the village of Royston and the aerodrome of Duxford from which the flights were made, this instrument was used throughout. It is noticeable that both the regularity and straightness of the strips of photographs, and the final accuracy of the mosaic, are greater in the latter than in the former half. Since the flying for these two halves was carried out on successive days, under identical weather conditions, by the same crew, using, apart from the rudder control, identical equipment, the complete mosaic affords good evidence as to the improvement in the work that results from using this rudder control. It is to be noted that the pilot found the flying for the Southern half much less fatiguing than for the Northern half; he attributes this entirely to the use of the rudder control. The gyro azimuth indicator was used in both flights.

We may summarise the data as to errors in the mosaic by saying that in the Southern half the probable error in the position of any point is about 0·02 mile and the maximum error about 0·05 mile, whilst in the Northern half the probable error is about 0·04 mile and the maximum 0·07 mile. These figures are definitely in favour of the part over which the gyro rudder control was used, in spite of the fact that this part of the ground was distinctly more hilly than the Northern part.

One other point of interest in the mosaic lies in the comparative absence of discrepancies at the joins between the separate prints, which, as we have seen, are due partly to errors in height and tilt, partly to changes of level in the ground, and partly to errors of compilation. It will be noticed that nowhere in the mosaic are there any errors of this nature much greater than the width of a road, so that this type of error rarely exceeds 0·01 mile.

The country of which the mosaic was made cannot be described as hilly, but it is not by any means flat. In several places there are moderately rapid changes of height of 200 ft. and more, and along the South edge there is a definite range of hills, rising to as much as 400 ft. above the lowest ground included in the mosaic.

During the compilation it was found that the Southern half was much easier to fit together than the Northern half, on account of the fact that, in the former, the flights had been straighter and the constancy of tilt and height had apparently been better maintained. This, no doubt, was due to the pilot having had more attention to spare for these operations, he being relieved by the rudder control from the necessity of keeping a straight course.

It was found, however, in the Southern half, that the part on the slope of the range of hills was more difficult to compile than the rest, owing mainly to the presence of strong fictitious curvatures, which had to be eliminated during the compilation. We consider, as the result of our experience on this mosaic, that moderately accurate mosaic maps could be made of country which is considerably more hilly than this, but it is clear that in hilly country the compilation will be more difficult than in flat country. We consider that the present order of accuracy could, with care, be maintained, provided that the local variations of height do not exceed about 5 per cent. of the height of the aeroplane, and that accurate mapping by this process would become impracticable when the variations of height exceed some 10 per cent. of the aeroplane's height, but on the latter point we have no direct evidence. Between these limits it is probable that useful maps could be made, but they would be neither so accurate nor so easy to compile as the specimen under consideration.

The fact that a mosaic of this size can be constructed to four control points only, and have an accuracy so great as that we have been discussing, came as a great surprise to us. We had expected to obtain moderate accuracy, as the result of the great care we had taken in developing the methods by which the photographs can be taken from a constant height and with the camera axis vertical, but we never expected to secure such accuracy as that recorded above. The independent results for the two uncontrolled mosaics combined with the result for the single large controlled mosaic show, however, that this order of accuracy is not accidental, but something definitely associated with the process employed. The accuracy of location of points far from control points depends, of course, entirely upon the successive fitting together of the detail at the edges of individual photographs; this accuracy will, therefore, fall off, should the individual photographs be distorted by tilt or by other causes. We have here, therefore, one important reason why the flying itself must be very accurate if good results are to be obtained; we have an example of the effects on general accuracy of a slight falling off in accuracy of flying in the comparison of the Northern and Southern halves of the mosaic.

Even given good flying and accurate photographs, we still do not think that anything like this accuracy could be realised unless the compilation were carried out by some systematic plan equivalent to that described. Preliminary attempts which we made to compile, without the elastic aids described, were quite unsuccessful from the point of view of accuracy or even of providing a continuous map. It seemed impossible during these attempts to avoid trying to get too good an agreement in some parts of the mosaic at the expense of parts not yet compiled. A dozen or more prints would be compiled, apparently successfully, only to find that the next print could not be made to fit at all, without disturbing the whole of the prints that had already been laid down. This was no doubt due to accumulated errors occurring of the kind that are so conspicuously absent from the interior of the mosaics which we afterwards constructed. The systematic methods, using elastic bands, allow the whole mosaic to be roughly erected first and then subjected to small proportional adjustments, which can be carried out whilst the whole surface is under observation, so that the danger of seeking too great accuracy in one area at the expense of another can be avoided. It is this general adjustment and averaging of the errors in the prints themselves which results in widely separated points being located in their correct relative positions with such high accuracy.

It should be noted that the systematic process that we have adopted would be almost impossible to carry out successfully unless the actual track of the aeroplane were approximately straight, so that the strips of prints do not have to be pulled much out of line to get them to their final positions. There is here, then, yet another reason why it is essential to have good straight flying, if the work is to proceed rapidly and accurate results are to be obtained.

Apart from the final accuracy of the finished mosaic, there are several other features of interest. The whole mosaic, covering 225 square miles of ground, was virtually constructed from the result of two flights, each of about three hours' duration from ground to ground; the time actually employed in photography was little more than two and one-half hours altogether. These flights were carried out on successive days and, from our experience, there would seem to be no reason why, given suitable weather, this rate of work should not be continued for many days at a time. This would lead to a rate of working of about 100 square miles to a flying day.

It is not precisely correct to say that the mosaic was constructed from the results of two flights only, because the camera shutter failed for three whole strips in the Southern part. It is known that the aeroplane flew over these strips with the same order of accuracy as the other strips for which photographic results were obtained, but the plates corresponding to them were blank. The gaps thus left in the mosaic have been filled in from a previous flight which had been carried out over the same area, but at a different altitude. The photographs of this flight were on a different scale from those in the rest of the mosaic, and they were, therefore, given a uniform increase of scale, the required ratio of increase being found by comparing one of the interpolated strips with one of the strips of the main mosaic that overlapped it. The strips which have been interpolated in this way are the first, the second and the fifth from the Southern edge. These interpolated photographs, being all reduced by the same ratio, retain all the small errors that are due to tilt and irregularities of height; although they are not contact prints, they contain all the errors of contact prints, and are equivalent to contact prints in so far as the compilation of the mosaic is concerned. Whilst, therefore, it is not strictly true to say that the whole mosaic was constructed from the result of two flights and from contact prints, the same deductions can be drawn from its results as if this had actually occurred; for the whole ground was covered in two flights and the prints all contain the errors appropriate to contact prints. All the remaining prints, other than those in the interpolated strips, are contacts from the original negatives.

The focal plane shutter of the L.B. camera, which we used, caused us endless trouble; its failures entirely spoiled several flights which we attempted, and were responsible for the three gaps in other parts of the mosaic, where single prints are obviously missing. There is no doubt that this design of camera is not really satisfactory for routine work, in which it is vitally necessary that there should be no failure, but it is understood that it is becoming obsolete. It is impossible to lay too much stress upon the importance of being equipped with apparatus in which failure is practically eliminated. The whole of the profits of a commercial venture in aerial surveying could very easily be wiped out by a few failures, necessitating an undue number of repeat flights.

Another feature of the mosaic that should be noticed is the remarkable straightness of the strips of photographs, particularly in the Southern half where the gyro rudder control was used. The bulk of the strips are not only very straight but very exactly parallel to one another. The straightness of the strips of the Northern part is a tribute to the skill of the pilot, flying officer Allan; that of the Southern half to the gyro rudder control. It is of interest to observe that the first strip on the first day, near the middle of the mosaic, is very badly out in regard to its direction. This was due to an error in calculating the compass course required to make good the desired direction over the ground. It must not, however, be imagined that such a large error is typical of the error to be expected from the method employed for finding the wind; it was in this case due to a definite defect in the compass, which was afterwards located. The compass was found to have been filled with dilute, instead of absolute, alcohol; this dilute fluid has a higher viscosity than absolute alcohol, so that the errors due to swirl in the liquid were accentuated, with the result that the compass took about twice as long as was expected to settle down after a turn. The observations required to determine the wind were therefore taken whilst the compass was still in error, with the consequence that the wrong result was reached. It is a point of considerable interest that, in spite of this large initial error, which we have since found can be avoided, the crew were able to correct the direction of flight after their first strip, so that the remaining strips lay, as accurately as could be wished, in the correct direction.

The photographic technique of the prints from which this mosaic has been compiled is exceptionally bad, and must not be considered in any way representative of what can be done in this direction. We were not ourselves investigating the photographic side of the problem, and we therefore used any material that came to hand which would give sufficient definition for our purpose. The plates from which these prints were made had been in the air on several occasions when, for various reasons, they had not been used; it is well known that this causes them to deteriorate rapidly, probably owing to damp being introduced into the plate box with the air that is forced into the box on descending from a rarefied atmosphere. It had been our intention to make another mosaic, embodying the results of our experience and using good photographic material, which would produce results more suitable for publication. We also wanted to have our show mosaic of the size recommended for routine work, namely 10 miles square, and to have it taken in one day, without gaps, and with all the strips lying in the right direction. We did in fact stand by for several months to get this show mosaic, but were unable to do so, owing to the weather. This gives some idea of the difficulty of contending with English weather, in work of this nature.

Even after Griffiths had left Cambridge we had intended to try and get a show mosaic made with him as observer, but his death made this impossible. I was unable, myself, to give the time necessary to practise the operations, and to stand by continually in the hope of catching any spell of fine weather that might occur. I had therefore to give up the idea of obtaining a really perfect mosaic for reproduction with this book, and to use our original experimental one, which is probably as accurate as any we should ever get, but which contains the numerous defects that have been discussed. These very defects have, however, an interest in themselves and it is probable that the only real harm that will result from our failure to obtain a perfect specimen of a mosaic is that readers will receive a bad first impression of the photographic technique when, as no doubt they will, they examine the map itself before reading the book.

We may summarise, in a very few words, the results and lessons to be learned from these experiments upon mosaic mapping:

1. It is possible to make good mosaic maps at the rate of 100 square miles to a flight of three hours' duration from ground to ground.

2. It is possible to construct, from the photographs so obtained, mosaic maps of about 10 miles square which will fit a true map with errors that average about 0·03 mile and seldom exceed 0·06 mile; these mosaic maps can be strained to fit arbitrarily selected control points.

3. Local discrepancies in the mosaic, which occur at the edges of the photographs, need not exceed 0·01 mile, so that, in circumstances in which control points can be provided at the rate of one per photograph, or about one per square mile, the absolute errors should rarely exceed 0·01 mile.

4. These results can be obtained in country in which local differences of level do not exceed some 5 per cent. of the height of the aeroplane. Useful, though not such accurate, results might be obtained in somewhat more hilly country.

5. To obtain results at this speed, and of this order of accuracy, it is absolutely essential to employ trained crews, suitably equipped and working on a carefully thought-out plan. But we think that any good pilot and observer could be trained to carry out these operations in a reasonably short time (see chapter VIII). It is also necessary to employ systematic methods of compilation, in which the distribution of errors is carried out by mechanical means, such as that provided by the elastic bands described.

There remains to be considered the problems of joining up the separate mosaics into a single large map, and of locating the true positions of mosaics that do not contain any ground-surveyed control points. We shall defer the consideration of these problems to the next chapter.

CHAPTER VI

MAPPING BY NAVIGATION AND OBLIQUE
PHOTOGRAPHS

IN the previous chapter we dealt with the construction of a single mosaic map of about 100 square miles in area; such an area will form only a very small part of the majority of surveying schemes in which it may be economically practicable to use aeroplanes, so that it becomes necessary to study the subject from a wider point of view. We have, for example, to arrange that the "indication"* strips, which may be required before the flying for the mosaics themselves can be started, shall be provided in an economical manner. The mosaic flying cannot be carried out in the same flight as that on which the indication strips are made, and the indication strips for a single mosaic do not provide enough work for a whole flight; some scheme must therefore be devised for making, in a single flight, indication strips for a number of mosaics. Again, we saw in the preceding chapter that the provision of independent control on the shape and scale of the mosaic merely caused a slight improvement in the general accuracy, and appeared as a matter of minor importance. From the wider outlook of a large survey, on the other hand, this question of control becomes of vital importance, for its absence would produce two disastrous effects; in the first place those mosaics which are far from any control might be bodily displaced from their true position, and in the second place adjacent mosaics might be slightly distorted so that they would not fit each other accurately at the edges. Whenever the ground control is sufficiently close for every mosaic to contain several control points, then each mosaic can be located in position and all can be strained, as in chapter v, to fit the control points and so eliminate systematic distortion. It is to be presumed that whenever practicable such a ground control will be provided, whereupon the problems under discussion in this chapter, in so far as they relate to the preliminaries of mosaic mapping, reduce to the simple matter of arranging for a series of more or less straight indication strips, which will themselves have no other function than to allow the pilot to locate the beginnings and ends of the mosaic flights. In these circumstances, the only point that need be mentioned here is that whenever the choice of position of control points is at the disposal of the surveyors, they should be placed near the corners of the individual mosaics.

* For the meaning of this term see chapter v, p. 48.

Although it seems possible that much of the mosaic work of the future will be carried out upon an adequate ground control, which will ensure that several control points occur in each mosaic, yet it is important that the aerial method should not be restricted, of necessity, to surveys in which such a control can be provided. There are types of country, for example heavily wooded plains and deltas, in which the mosaic method may reach its greatest value, but in which provision of control points by triangulation would be impossible, except at prohibitive cost. In the interior of such countries the location of the mosaics themselves and the joining together of the separate mosaics will present serious difficulties.

Again, when the aerial method is used in connection with any extensive survey, there will be many areas to which the mosaic method is not applicable, for two reasons: the ground may be too hilly, or it may be of insufficient interest to warrant the cost of the mosaic method, which requires about one mile of survey flying for every square mile of country mapped. The areas which come into these two classes may indeed greatly exceed the areas which are really suitable for mosaic mapping. Hilly country, of sufficient interest to warrant a really close ground survey, such as will provide suitable control points spaced not more than one or two miles apart, comes into the class suited to the methods of re-section, a branch of aerial surveying with which this book does not deal. Much of the hilly country of the world will not, however, in the early stages of development, be of sufficient interest to warrant such a close ground survey, or the expense of detailed re-section. There is thus a need for some method or methods suitable either for hilly ground which does not warrant the expense of re-section, or for flat ground which does not warrant even the expense of the mosaic method.

It might be thought that difficulty will also be experienced in constructing mosaics of areas, such as deserts and forests, in which there are no easily recognisable features, but experience in the Sinai desert during the war showed that ground, which is apparently very featureless, may contain differences of shade appearing on the photographs with sufficient clearness to allow the mosaic to be constructed.

We have given considerable thought to these two problems, on the one hand of providing aerial control upon mosaic mapping and on the other of making cheap maps of those classes of country to which the mosaic method is not applicable, and we have come to the conclusion that the basis of a solution to both problems lies in what we now call Navigational Control. The primary idea of the process, which we call by this name, is that the position of the aeroplane shall be determined by the direction and speed of flight and the time since passing over a control point; but it may be stated at once that the practice of aerial navigation has not yet reached the stage in which direct calculation, based on the time since passing a single control point, will locate the aeroplane with an accuracy sufficient to be of much value in any but the very roughest form of surveying. This being so, the method of applying navigational control which we have followed assumes that points near the beginning and end of a long straight flight have been independently located, so that it is only necessary to keep the flight straight and at a constant speed to be able to locate, by timing, the position of the aeroplane at any intermediate instant.

The method by which the beginning and end of a long flight may be located will vary according to circumstances; it must always be based upon some form of ground control, but since the straight flight may be anything up to a 100 or more miles in length, the method is applicable when only a very open ground control is available. We started to study this method of navigational control purely with the idea of providing control over a large map constructed from many mosaics, when the ground control alone is inadequate; but we soon realised that, combined with the use of oblique photographs, the method provides a very hopeful way of dealing with the two classes of country mentioned as falling outside the scope of the mosaic method. We, therefore, continued our experiments with a view, not only to the provision of control upon mosaics, but also to the development of methods which would produce, rapidly and cheaply, approximate maps of country that may be either unsuitable for mosaic surveying, or of insufficient interest to warrant the time and expense per square mile of the mosaic method.

When required merely as an adjunct to mosaic mapping, navigational control could obviously be applied most conveniently by flying the indication strips as straight and steadily as possible, and arranging that they should be independently located at their ends, but once the idea of reconnaissance mapping by obliques was formed, it became clear that it would often pay to take oblique photographs simultaneously with the verticals of the indication strips; for these provide valuable information about the nature of the ground which eventually is to be mapped by verticals. This information, in the first place, might enable the director of the survey to decide which areas of the ground are of too little interest to warrant proceeding to the mosaic method, in the second place it should greatly assist the pilot in identifying the ground to be surveyed, and in the third place it could be used to increase the accuracy of the final mosaic, because straight lines in the obliques will also be straight lines in the mosaics.

For these reasons it would seem that the best way of beginning an aerial survey of any large unmapped country would be to carry out a series of long straight and parallel flights, spaced at convenient intervals, say ten miles apart, and during these to take at frequent intervals both vertical and oblique photographs. The data so obtained might either be worked up into what may be termed a reconnaissance map, or be used as the basis of a subsequent detailed mosaic survey.

The value of such a reconnaissance survey depends upon the accuracy that can be obtained from it with various kinds of ground control, upon the amount of detailed information that it will provide, and upon the ease of construction of the map from the oblique photographs. It was to throw light upon these points that the experiments to be described were made. Obviously the precise nature of the scheme to be adopted and the order of accuracy to be expected will vary greatly with the nature of the ground control available. It may sometimes be possible to "traverse" on the ground along the path of the aeroplane, and fix large numbers of points occurring on either side of the path; at other times, it may be necessary to span large areas of ground containing no control points, and to rest the ends of the straight flights upon strips of known country separated by hundreds of miles.

The method used will have to be suited to the particular problem in hand, and it was obviously impracticable for us to attempt a detailed trial and investigation of methods suitable for all possible conditions which might arise. Our resources in fact would not allow us even to investigate one set of conditions thoroughly, so as to provide conclusive proof of the order of accuracy obtainable; such for example as we were able to give in the case of mosaic mapping. The best that we could hope to do was to carry out a series of trials which would provide data from which the probable value of a map made by these methods might be estimated for any given set of conditions. Even this we have not been able to do so well as we would have wished owing to the fact that both Griffiths and his pilot, flying officer C. Allan, had to abandon the work soon after it was started.

Some data has, however, been obtained, both before and after the departure of these officers, and, though it is not quite so complete as we had hoped, it should be sufficient to allow a fairly reliable estimate of the probable value of the method to be made. It should be realised, however, that the accuracies which we shall discuss in this chapter are almost certainly lower than are to be expected from pilots engaged in the work as a regular routine. Our experiments have been so hampered by the unfavourable English weather that we have been able to carry out flights at long intervals only, so that the experience gained in one flight is almost forgotten before the next occurs; moreover, since

Griffiths' departure, the flying has been done in occasional spare moments snatched from other routine work, with the consequence that neither the pilot nor the observer have been able to attain that degree of familiarity with the routine of the operations which is essential to obtain the highest accuracy in aerial, as well as other experimental work. It may therefore be taken that the accuracies to be discussed will almost certainly be exceeded by any body of trained pilots and observers, using the same method and working regularly to a routine, which provides, for the time being, their whole occupation.

The two problems of applying navigational control and working quickly and cheaply from oblique photographs, although they are ultimately to be associated, are really quite distinct. Let us begin with the problem of navigational control.

The idea is to make long straight and steady flights, in which the position of the aeroplane at the beginning and end shall be known, and to estimate the position of the aeroplane, when at intermediate points, on the assumption that the flight has been straight and the speed constant throughout. The accuracy with which this process can be carried out depends upon the straightness of the path and the constancy of the speed of the aeroplane *over the ground*. Errors may be introduced from three causes: wind changes during the flight; inaccurate flying; inaccurate timing. Errors due to the third source should be negligible in comparison with those due to the other two.

Let us deal first with errors due to wind changes, on the assumption that the flying and timing have been accurate and that the ends of the flight have been accurately located. Any change of wind during the flight will cause the path over the ground to be curved and the speed to change, with the result that the calculated position of the aeroplane at any spot intermediate between the ends will be displaced from the true position by an amount, the magnitude and direction of which we shall describe as the "vector error."*

It can easily be shown† that a wind, changing steadily with time during the whole flight, will cause vector errors which will reach a maximum half-way through the flight by time, and that this maximum vector error will be one-eighth the product of the time of flight and the vector change of wind during the flight‡. For example, in a flight of half an hour's duration, during which the vector change of wind was four miles an

---

* If a point situated at *A* is thought to be situated at *B*, the line *AB* which displays both the direction and amount of the error is defined as the "vector error."

† See Appendix III.

‡ If a wind that can be represented in magnitude and direction by a line *AB* changes to one that can be represented by a line *AC*, then the line *BC* is defined as the vector change of the wind.

hour, points near the middle of the flight would be located with an error of one-quarter of a mile.

Apart altogether from our own experiments, there is a certain amount of data available relating to the changes of wind which occurred with lapse of time and change of position in certain regions of the British Isles and France. It is of some interest, before proceeding to discuss our own experiments, to make an estimate of the errors that this information would lead us to expect, on the assumption that the pilot flies, *through the air*, in a perfectly straight line and at a constant speed. The most reliable summary of these data is probably that given by Dobson in the *Aeronautical Journal* for May, 1921, p. 226. This data is shown, in Appendix III, to lead us to expect that the 50 per cent. error* near the middle of a navigational flight of this kind should be about 0·7 per cent. of the distance between control points when this distance is of the order of 50 miles, and about 1 per cent. of the distance when it is in the neighbourhood of 100 miles.

With a vector quantity of this sort we should expect the error that will be exceeded once in every ten occasions to be 1·8† times as large as the 50 per cent. error, which is exceeded every other time; hence we may anticipate occasional errors of 1·3 per cent. or 0·6 of a mile on a flight of 50 miles, and 1·8 per cent. or 1·8 miles on flights of about 100 miles in length. These figures take no account of the pilot's flying errors; it will be found of interest to compare them with the errors actually observed in the experimental flights to be discussed later.

In order to obtain direct evidence about the accuracy given by the method of navigational control we undertook a series of flights, varying from about 40 to 90 miles in length, in which the pilot flew through the air as straight and as steadily as he could, whilst photographs were taken at known intervals by a "vertical" camera. The optical centres of these photographs were then identified and plotted on existing maps and compared with calculated positions, obtained on the assumption that the first and last photographs of each flight had been accurately located, and that the aeroplane had passed over the ground in a straight line and at a constant speed during the whole flight. The majority of these flights are shown plotted in Diagram 16, in which the true positions of the optical centres are shown as dots, and the calculated positions by the intersections of long straight lines with short transverse lines, spaced according to the recorded times at which the photographs were taken. The distances between the dots and the intersections are, of course, the errors that

---

* The 50 per cent. error is that error which is equally likely to exceed, or be exceeded by, any individual error; see Appendix I.

† See Appendix I, p. 141.

would have been given by the process, had there been no independent maps of the ground between the beginning and end of the flight, and had each flight been considered by itself, without reference to any of the other flights by which it is intersected.

When a number of intersecting flights such as those in Diagram 16 cross an unknown area, it is possible, by considering them together, to obtain more accurate results than could be got from the same flights considered separately; the process by which this can be done will be considered later, but for the present we shall study the errors given by the individual flights standing alone. For this purpose the scale on which it has been possible to reproduce the flights is much too small, but an analysis of the original data has been made in the case of nine flights, six of

Table 1.

Showing flights in experiments on Navigational Control.

Pilots :   A. = C. E. H. Allan.   B. = D. L. Blackford.
Observers :   G. = J. C. Griffiths.   J. = B. M. Jones.

Distances given in miles. Angles in degrees.

Flight	Pilot	Observer	Date	Length	Intended direction (True)	Wind strength M.P.H.	Wind direction (True)	Drift angle	Average direction error	Gyro Rudder Control
G	A.	G.	18. xii. 22	63	182	40	252	23	2	—
D	A.	G.	12. vi. 23	63	182	26	326	9	0	—
H	A.	G.	12. vi. 23	45	2	26	326	9	2	—
A	A.	G.	29. vi. 23	90	92	19	—	10	6	—
B	B.	J.	6. vii. 23	92	92	14	167	9	$\frac{3}{4}$	g
F	B.	J.	9. viii. 23	56	182	48	279	30	2	g
C	B.	J.	10. viii. 23	65	182	32	274	19	4	g
E	B.	J.	10. viii. 23	54	92	32	274	1	$\frac{1}{2}$	g
K	B.	?	11. i. 24	37	182	—	—	—	—	—

which are reproduced in the diagram. A schedule of these flights together with certain relevant data is given in Table 1. In this table the flights are arranged in chronological order and are given distinguishing letters, by which they can be identified in the diagrams and in the following discussion. The direction and strength of the wind, as determined in each case just before the flight, at the survey height, are given, in order to show the conditions under which the flight was made, and the drift angle corresponding to this wind is given also. It will be observed that some of the flights were carried out in strong cross winds which necessitated flying with large drift angles. The last but one column shows the angle made by the line joining the beginning and end of each flight with the direction in which it was intended that the flight should take place*. Except in two cases angles are given to the nearest degree only and speeds to the nearest mile per hour.

The last column shows whether the gyro rudder control was, or was not, used. Those flights in which it was used are marked "*g*"; in the other flights it was out of action. This apparatus did not work perfectly in any of the flights here recorded; it precessed continuously, owing apparently to its having been slightly damaged during an accident, earlier in the year, in which the aeroplane was wrecked. The pilot counteracted this slow precession by occasional movements of the differential adjustment, which allowed the relation of the aeroplane to the gyro wheel to be altered without disturbance to the wheel itself. The gyro azimuth indicator was available in all the experiments.

Besides the flights in this schedule, several other flights which were made were unsuccessful, owing to causes such as the formation of clouds, the failure of instruments, or of the camera; the schedule contains, however, all flights in which definite failure of this sort did not occur, so that the results of the flights now to be discussed are not in any way selected.

Three of the flights in the schedule do not appear in Diagram 16. Of these *G* was the first flight of this kind attempted and was carried out before the final form of the scheme of Diagram 16 had been settled, with the consequence that the distribution of the photographs in this flight was not suitable to the rest of the scheme. This flight was therefore repeated by flight *C* in which, as will be seen later, the errors are much of the same order as in *G*. Flight *H* was intended to occupy the position of the strip marked *E* in Diagram 16, but after 45 miles the pilot, who had already carried out flight *D*, was taken with cramp in one leg, and whilst readjusting his feet, lost his line completely. The flight has therefore only been analysed for the first 45 miles, although it actually continued for over 60 miles, with the last 20 miles displaced sideways by more than a mile. The pilot recorded this difficulty with cramp before the results of the flight had been ascertained. The place of this flight has been taken by the flight now marked *E*, which was actually flown in another place and in another direction altogether. This flight has been transplanted bodily to its present position by the help of transparent paper, but the original error in direction, in this case amounting to only one degree, has been carefully preserved. With the exception of this flight *E*, all the flights in Diagram 16 are shown exactly as they fell upon the map, so that the amount by which the pilots failed to carry out the

---

* The reason that the intended direction of flight was always two degrees removed from the cardinal points is that it was desired to have these directions parallel to the edges of the maps upon which the comparisons would be made and, owing to the system of projection used in constructing the Ordnance maps, their edges do not lie true North and South.

prearranged scheme, involving flights spaced 20 miles apart and parallel to the sides of the map, is obvious on inspection. Finally flight K, which failed after 37 miles, owing to failure of the camera, was an attempt to fill the place now occupied by E. The observer in this flight was inexperienced and no attempt was made to find the wind, the direction of the flight being found by guessing.

The last column but one of Table 1 is interesting; it shows that out of eight flights in which the wind was determined and the required bearing calculated, in only two was the error of the average course over the ground greater than two degrees, although on several occasions strong cross winds were present. The cause of the large error of six degrees in flight A is not known; it was not discovered until after Griffiths had left the work and his death made it impossible to enquire the cause; it unfortunately happened that the notes which he left on this flight were incomplete, owing apparently to a page having been blown overboard. It is probable that an error occurred in the reduction of the results. The other large error of direction, that of C, is traceable, in part, to changes of the aeroplane height in a wind which was varying rapidly with height; this point will be reconsidered later. The results in this column suggest that, with careful work, the average direction of a flight should lie within some two degrees of the desired direction, but that occasional errors greater than this may occur, either owing to mis-reading of the instruments, or to an unfortunate combination of initial errors superimposed upon a change of wind.

We may now turn to a discussion of the results which these experiments were primarily intended to provide, namely data concerning the accuracy with which intermediate points on the flights can be located, when the positions of the ends of the flights are known. A distinction must here be drawn between what will in future be described as "local" errors and "total" errors. By local errors we mean the displacement of a point in a map relative to its immediate surroundings, whilst by total errors we mean the displacement of a point relative to accurately located control points, which may be either near, or far, from the point under consideration. Thus, in these flights, the displacement of any photograph relative to the adjacent photographs in the series will be classed as a local error, whilst its displacement relative to the supposedly known beginning and end of the flight will be called the total error.

In the present connection the distinction between these two aspects of the errors in a map is an important one. Where the accurately known control points are few and widely spaced, it may be essential to keep the local errors down so as to obtain a true map of any limited district, although relatively large errors may be admissible in the location of that district in relation to distant control points. It should be noted that, in the method now under consideration, local errors will not be dependent upon the spacing between control points, unless these are so close that at least one appears on each photograph, whereas total errors will probably depend directly upon the distance between the control points. When, therefore, the local errors which are associated with the method exceed the local errors that can be accepted, the method is not suitable for the purpose in hand, and nothing short of a very close control, the necessity for which might entirely alter the character of the survey, could enable satisfactory results to be obtained. When, on the other hand, the local accuracy obtainable is sufficient for the purpose in hand, then the data relating to total errors may be considered as providing information upon which a suitable spacing of the ground control, to give any required accuracy, can be estimated.

Table 2. *Local Errors.*

Distances in miles.

Flight	Lateral Components		Fore-and-aft Components		Approximate distance between consecutive photographs
	50 % error	Maximum error	50 % error	Maximum error	
A	0·06	0·15	0·04	0·12	4
B	0·03	0·12	0·07	0·20	3
C	0·02	0·07	0·03	0·08	$4\frac{1}{2}$
D	0·02	0·07	0·02	0·07	2
E	0·02	0·08	0·03	0·10	$4\frac{1}{2}$
F	0·02	0·08	0·04	0·08	$3\frac{1}{2}$
G	0·01	0·03	0·02	0·04	$1\frac{1}{2}$
H	0·04	0·11	0·03	0·09	$3\frac{1}{2}$
K	0·02	0·08	0·03	0·13	4

Let us first study the local errors of these flights. This has been done by measuring the displacement of each optical centre from the point that lies mid-way between the optical centres of adjacent photographs before and behind it. Had the flights been straight and at a constant speed throughout, and had the time intervals been all equal and the camera axis always vertical, displacements so measured would be zero; hence the measured values of these displacements will indicate the amount by which the assumed position of each photograph will be in error, in relation to its immediate surroundings*.

* In certain cases the time intervals, though accurately known, were not equal, and in these cases, which are comparatively rare, the local errors have either not been calculated, or appropriate allowances have been made. Such occurrences are due either to forgetfulness on the part of the observer, or to time lost in changing boxes of plates; they introduce no serious difficulty into the analysis of the flights, but merely complicate it slightly and disturb the even appearance of diagrams, such as 16.

In Table 2 the local errors have been separated into components, parallel and perpendicular respectively to the direction of flight. The 50 per cent. and the maximum values of these components are given for each flight. The magnitude of the 50 per cent. local errors provides a good measure of the quality of the flying in the flight to which it refers.

It will be noticed that the largest errors occur in flights *A* and *B* and it happens that these large errors are easily accounted for; in flight *A* the pilot complained of not feeling well and the flight would not have taken place at all, had not Griffiths been very anxious to complete one more flight before leaving for other employment. The same crew had previously carried out flights *G, D* and *H*, with results which are much more accurate from the present point of view. Flight *B* was the first attempt of a new crew, and affords a good illustration of the statement made in a later chapter on the training of personnel, that a pilot must be shown the errors which have resulted from small lapses of attention during his flight, before he can be expected to do his best work. Flights *F, C, E* and *K* were made subsequently by this same pilot, after further practice, and these serious errors do not recur.

Omitting flights *A* and *B* as not being representative of normal conditions, we can obtain some useful information from the data in this table. In the experiments of chapter IV we found that the probable value of a component of tilt about any given axis was in the neighbourhood of three-quarters of a degree, and it is easy to show that random tilts of this magnitude would account for the major part of the local errors observed in the flights now under consideration, even on the supposition that the aeroplane had passed over the ground in a perfectly straight line and at a constant speed. We may infer from this that the flying during the navigational work was not worse, as regards tilt, than that analysed in chapter IV, in spite of the fact that the navigational flights were very much longer than the earlier flights, and that more attention was being given to the maintenance of a correct compass course and a constant speed. We may also infer that the local errors are due in a large measure to accidental tilts and hence would have been reduced had a knowledge of the tilts allowed the plumb points to be determined. This last remark may be put differently by stating that the passage of the aeroplane over the ground was probably straighter and occurred at a more constant speed than the optical centres of the photographs would lead us to suppose. Some support is lent to this last inference by the fact that the local errors do not depend to any great extent upon the distances between consecutive photographs, as would almost certainly be the case were they caused mainly by changes of direction or speed.

The conclusion that a large part of the local errors was due merely to unknown tilts of the camera is one of some importance, for if it is true, it follows that the quality of the work could be improved if the tilts were known, so that the true plumb points could be ascertained. In the flights of chapter IV it will be remembered that the tilts were measured and the actual plumb points plotted; these flights have, therefore, been analysed for local errors, in the same way as the navigational flights, and it has been found that the 50 per cent. local lateral error of the plumb points is only about 0·01 mile as opposed to 0·02 to 0·04 mile found for the optical centres of the navigational flights. It should, therefore, be safe to assume that the local errors of a series of true plumb points will be considerably less than the local errors of a series of optical centres such as those in the navigation experiments*. We shall return to this point when we consider, at a later stage, the advantages of including some part of the horizon in one or more of the cameras carried in a survey flight.

One other point of interest that arises from the data contained in Table 2 is that, as regards local errors, the results obtained from the two pilots are practically identical. This supports the opinion, several times expressed, that similar results could be obtained from any good pilot after suitable training.

The data relating to local errors may be summarised by the statement that the 50 per cent. value of both the lateral and fore-and-aft components of the local errors should lie between 0·02 and 0·04 mile when there is no means of determining the accidental tilt of the camera, but that they should be more nearly in the neighbourhood of 0·01 mile when the tilt is accurately known. With inexperienced pilots, or with pilots who are not in good condition, the errors may be greater than this.

It is of some interest to know the accuracy with which the pilot has managed to keep his aeroplane flying on a constant bearing during the flights. The azimuth orientation of the photographs can be determined

---

* It is interesting, in this connection, to note that, if an aeroplane at height H flew in a path that was a true sine curve of periodic time T, about a mean straight line, and if the aeroplane were always correctly "banked," then the displacement of the optical centre of a vertical camera would always be $\left[1 + \dfrac{4\pi^2}{T^2} \dfrac{H}{g} \right]$ times the displacement of the aeroplane itself. For a height of 10,000 ft. this factor would have the value 2 when

$$T = 2\pi \sqrt{\frac{H}{g}} = 110 \text{ seconds.}$$

In actual practice the aeroplane is not always correctly banked so that this result must not be given too much weight in connection with practical work, but it helps to explain why the displacements of the optical centres of a series of photographs from a mean straight line is, in general, greater than the displacements of the aeroplane itself.

by comparing them with the map and, since the camera remained fixed in the aeroplane during any flight, the variations in the bearings of the aeroplane, at a number of instants during the flights, can be ascertained. The pilot of course was attempting to maintain a constant bearing during each flight, so that the deviations of the instantaneous bearings from the mean bearing for the whole flight give a measure of the accuracy with which he has succeeded. In Table 3 are given the probable deviations and the maximum deviation from the mean bearing of all the flights.

Table 3. *Deviations in the bearing of the aeroplane from the mean bearing for each flight.*

Deviations in degrees.

Flight	No. of photographs	50 % deviation	Maximum deviation
A	39	1·6	8·6
B	52	1·3	9·8
C	25	0·6	1·7
D	20	0·8	2·5
E	22	0·8	3·1
F	29	0·7	3·5
G	23	1·3	3·6
H	16	1·7	7·0
K	11	0·8	1·9

It appears from this table that it is practicable to hold the aeroplane, even in a long flight, so that it never deviates more than some two or three degrees from the mean bearing for the flight, and to keep the 50 per cent. deviation down to something under one degree. It will be noticed that *A* and *B* are again amongst the bad flights and that one must expect occasional lapses amounting to eight or nine degrees, due no doubt to the pilot's attention being distracted, or to fatigue. The large deviations in *A* and *H* both occurred after the pilot had been flying for some hours, whilst that in *B* occurred right at the beginning of the flight and may have been due to the pilot not having settled down on his course. For future reference we may put the 50 per cent. deviation from the mean bearing of a flight at about one degree.

The flights in which the gyro rudder control was used are *B*, *C*, *E* and *F* and these are on the whole better than the others, but the results for *D* and *K* in which the rudder control was not used show that equally good results can be obtained without the gyro. It is to be remarked that the gyro was not working perfectly on any of these flights.

Let us now take up the study of the total errors, assuming first that each flight is considered by itself, without any reference to the other flights that cross it, and that the position of the optical centre of each photograph is determined on the assumption that the flight was carried out in a straight line and at a constant speed, between known points which occurred in the first and last photographs. These errors can be seen on a very small scale in Diagram 16, but, for the sake of convenience of discussion, Table 4 has been prepared showing the maximum errors that occurred in each flight. In this table are shown the maximum lateral errors, the maximum fore-and-aft errors and the maximum vector errors that occurred in each flight, and these are all given both as distances in miles and as percentages of the total length of the flight.

Table 4. *The Maximum Errors in the Navigational Experiments.*

Flight	Length Miles	Miles			% of distance between control points		
		Lateral	Fore-and-aft	Vector	Lateral	Fore-and-aft	Vector
A	90	0·45	0·80	1·02	0·50	0·89	1·13
B	92	0·57	0·42	0·62	0·62	0·45	0·67
C	65	0·50	0·75	0·80	0·77	1·15	1·23
D	63	0·40	0·17	0·42	0·63	0·27	0·67
E	54	0·17	0·47	0·50	0·31	0·87	0·93
F	56	1·10	0·35	1·15	1·96	0·62	2·05
G	63	0·82	0·32	0·90	1·30	0·51	1·42
H	45	0·25	0·23	0·27	0·55	0·51	0·60
K	37	0·18	0·53	0·53	0·49	1·43	1·43

An examination of this table shows that the maximum vector error in a flight has a 50 per cent. chance of exceeding 1 per cent. of the length of the flight, whilst the maximum vector error recorded in any of the nine flights is just over 2 per cent. If these figures are compared with those given earlier in this chapter, based on Dobson's figures for wind changes with space and time and obtained on the assumption of perfect flying, it will be seen that the 50 per cent. error actually found is equal to that which would be expected on a flight of 100 miles, and is not much greater than the figure given for a 50 mile flight. As most of the experimental flights were between 50 and 100 miles in length, the agreement is as good as could be expected and suggests that, in these experiments, the bulk of the errors were due to wind changes rather than to imperfections of flying. This suggestion is supported by the fact that the total errors in flights *A* and *B* are no worse than in other flights, although the local errors recorded in Table 2 show that the quality of the flying in these two flights was unusually low.

An examination of the figures for lateral and fore-and-aft errors shows the 50 per cent. value of both components of error to be 0·70 per cent. This striking equality in the two classes of error again suggests that the major cause of error is wind change; for if we are to attribute the errors to imperfect flying this equality would have to be assigned to a remarkable coincidence, since the causes that lead to errors of direction in flying are very different from those that lead to errors in speed.

The question of the extent to which the errors are due to wind changes on the one hand, or to defective flying on the other, is one of some importance; for, if they are mainly due to wind changes, then better results would be expected in parts of the world where the winds are less variable than in the British Isles, whereas if they are due to imperfect flying they might be reduced by the use of automatic apparatus for controlling the aeroplane, or by improvements in the sensitivity of the controls and in the training of the pilots. On the whole the evidence so far discussed is distinctly in favour of the supposition that the total errors were mainly due to wind changes and the local errors to imperfect flying.

If, in Diagram 16, we examine the individual flights, we find that the flight involving the greatest error, namely F, is peculiar in that it has a sharp bend about a third of its length from the start, the two portions before and after the bend being approximately straight. Table 3 indicates clearly that this bend was not due to the aeroplane turning off its course; for had it done so by an amount necessary to produce such a change of direction, the 50 per cent. deviation of the azimuth orientation from the mean for the flight could not have been so small as three-quarters of a degree. Indeed a comparison of the average bearing of the aeroplane in the eight photographs preceding the turn and in the twenty-one that were taken after the turn, showed a difference of less than one degree. From Table 1 it will be seen that the aeroplane was flying in a southerly direction in a west wind of some 30 miles an hour, and, since it did not turn or seriously alter speed, the only remaining cause that could have been responsible for the bend is a reduction of strength in the cross wind. It is in fact easily calculable that the wind acting on the aeroplane must have reduced in strength, more or less suddenly, by about 10 miles per hour, in order to have caused such a change in direction in the path of an aeroplane moving with an air speed of some 96 miles per hour.

An investigation into the height of the aeroplane during this flight has been made from a consideration of the scale of the photographs, and it was found that, after remaining in the neighbourhood of 10,600 ft. whilst the first four photographs were taken, the height fell suddenly, so that by the sixth photograph it was down to 10,250 ft., and, henceforward, remained within about 100 ft. of this value for the rest of the flight. The change in the strength of the wind could therefore be accounted for on the supposition that the wind was changing with height at a rate of over three miles per hour per 100 ft. This is a very great rate of change to occur at such a height as 10,000 ft., but the evidence for its existence is nearly conclusive.

The correlation of the very noticeable bend in flight *F* with a change of height of the aeroplane, led us to examine the other flight in which there was a strong cross wind. In this flight, *C*, the bearing of the aeroplane was particularly well maintained, as can be seen from Table 3, but an inspection of Diagram 16 shows that its path deviated to the west, after the fifth or sixth photograph, and returned later to the original line of flight. In this flight the height by aneroid was frequently recorded and it appears that a sudden rise of about 120 ft. occurred after the fifth photograph, and was maintained until the sixteenth, after which a fall of about 200 ft. occurred. An inspection of the course of the flight will show that these changes of height would account for the curvature of the course, on the assumption of a wind gradient giving, as in the case of flight *F*, a decreasing wind with increase of height.

In carrying out these navigation flights it was not considered to be necessary to lay great stress on the maintenance of a constant height, since this was thought to be a point of minor importance in relation to the particular problem in hand. It was not anticipated, however, that the pilot would make such a large change of height as is here recorded, nor was the presence of such extreme velocity gradients suspected. It is probable that, in concentrating on the maintenance of direction and speed, the pilot had sacrificed accuracy in height, with the result that he reached a different current of air from that in which he started and hence caused the bends recorded.

If we are correct in attributing the most serious error in these experiments to an accidental change of height, it becomes a matter of the highest importance to attend closely to the maintenance of a constant height, both during the navigation flights themselves, and while measuring the wind beforehand. This adds, of course, to the burden laid on the pilot, and affords an additional reason for the development of some instrument that will automatically keep the aeroplane at a fixed height. Reference to the need for some such instrument is made again in chapter VII.

We may summarise the data relating to total errors in individual flights by the statement that the 50 per cent. value of the maximum vector error in any flight has been found, in our experiments, to be 1 per cent. of the distance between the known points, that the evidence is in favour of the observed errors having been mainly due to wind changes, and that the largest error of all—2 per cent. of the length of the flight—was caused by an unintentional change of height, which carried the aeroplane into a new wind current.

These are errors given by isolated flights, without reference to their situation in any larger scheme. In an actual survey of any large unknown area it will generally be necessary to carry out a series of parallel flights, spanning the area, and spaced not more than some 10 miles apart; if the length of these flights between control points is much greater than their distance apart, certain difficulties will arise that have not yet been discussed. Consider two adjacent flights, some 50 miles in length between control points, and accidently bowed in opposite directions, so that, near the middle of each flight, the aeroplane was, in each case, some half a mile nearer to the other flight than would be supposed on the assumption that it had travelled in a straight line. A map based solely upon these flights would show the distance between the photographic strips near their centres as one mile greater than it really is, so that a distance of 10 miles would be subject to a local error of 10 per cent. Such an error might easily occur under the conditions imagined, and even larger errors would be expected when the ratio of the lengths of the flights to their lateral spacing is greater than here supposed. We have here then a very serious limitation upon the application of the method of navigational control, so long as each flight is merely considered by itself, without reference to adjacent flights.

Again, it is of the greatest importance, in any survey, to have some means of checking the errors that are occurring, because it is always possible for some complete mistake to occur, or for some exceptionally large error, such as that in strip *F*, to be introduced. When such a large error does occur, it is important to have some warning of its existence, so as to be able to repeat the relevant observations, should it be considered desirable to do so. Such a check is very difficult to obtain, when working by the method under discussion, so long as each flight is considered by itself, but it can be obtained if the flights are linked together in some way so that each acts as a check upon its neighbours.

There are thus two distinct reasons why it is important that the individual flights should not be treated as independent units, but should be linked together in some way to allow unusually large errors to be detected and to prevent the occurrence of large local errors. This linking up can be effected at the expense of a series of additional flights, which cross the main system of parallel flights and divide the country up into a net of intersecting strips of photographs. Let us suppose for a moment that this second series of strips of photographs begins and ends, like the primary strips, upon control points; then it will be clear that those areas of ground which appear in two strips, at their intersections,

6–2

will be located in two independent ways, and that each strip will provide a check upon the accuracy of the other. Again, consider two parallel strips crossed near their middle by a third strip lying approximately at right angles to them; if the intersection points can be identified upon the photographs of the strips concerned, we have at once a check upon the length and orientation of a line joining the first two strips; for we know the average speed and direction of the flight that crosses the other two, and the actual speed and direction during the time of crossing are not likely to differ from the average speed and direction for the whole flight by anything like so much as the 10 per cent. error, which we have seen might easily occur in the absence of the crossing flight.

Stating the matter more generally; when an unknown area of ground is covered by a net of intersecting flights, each of which provides data whereby any point which occurs in its photographs can be located with a given probability of error, then the whole data from all the flights will not be self-consistent, in that all intersection points will be located twice with different errors. The fact that the data is not self-consistent will be obvious to the surveyor, so that he can adjust the errors until the data is self-consistent, and it should be possible, during this adjustment, to bring about three separate improvements: the probability of error in the whole system should be materially reduced; the danger of excessive local errors of the type just discussed should be avoided; the existence of any isolated large error should be indicated.

The problem of how best to distribute the errors so as to achieve these three ends might be solved mathematically and the solution applied in practice by some process of computation, but the difficulties of this method of attack may be very great; for the probability of error in a given flight varies, not only with the length of the strips, but with the position of each point in its strip. Thus, an intersection that is near the end of one strip and far from either end of the other is much more accurately located by the former than by the latter, and allowance must be made for this difference in any adjustment. It happens fortunately that this problem of adjusting the errors so as to give each strip its due influence in the final adjustment can be solved approximately by a very simple mechanical process, which is based upon the fact that when an elastic string is stretched to an appropriate tension, between fixed points, the rigidity with which any point along its length is held in position by the tension and rigidity of the string varies along the length of the string in the same way as the improbability of error varies along the length of the flights. That this is approximately so can be seen by inspection of the problem; for, in the case of the string, points near the ends are very rigidly held in position compared with points near the middle; whilst, in the flights, the probability of error is small near the ends and large near the middle. A quantitative discussion of this comparison is given in Appendix III, Part 2, where it is shown that the comparison can be made very exact, if it is assumed that the errors in the flights are produced solely by a steady rate of change in the wind.

This mechanical distribution of errors can be effected in practice as follows: the supposed positions of the plumb points at the moment of exposure of each photograph are first set out as on the straight lines in Diagram 16, each plumb point being assumed to lie on the straight line joining the control points, and to divide this line strictly in proportion to the time intervals between its exposure and the exposures of the control photographs, which in this diagram were taken to be at the beginning and end of the flight. Elastic strings are then stretched between the control points to represent each flight, and are marked* at these plumb points. The exact times of crossing the other strips passed over during the flight are then worked out, and additional marks, spaced proportionally to these times, are placed on the elastic strings. This of course can only be done when some recognisable ground feature appears in the photograph of both strips, near where they cross. When this has been done to every strip there will be, for each intersection, two marks, purporting to represent the intersection point, but owing to errors neither of these will lie on the true intersection point. The elastic strings are then forcibly strained until these calculated intersection points do coincide, and are fastened together with seccotine in this position. The whole net is then allowed to take up its own formation and the new positions of the plumb points on the elastic strings are taken to be the best estimate that can be made of the true positions of the plumb points. This whole process is very rapid and can be carried out, for a scheme such as that indicated in Diagram 16, in a few hours.

The scheme of flights shown in Diagram 16 was arranged with the intention of testing, amongst other things, the method of elastic adjustment of errors that has just been described. These flights cover an area of about 90 miles by 50 miles, which may, for this purpose, be imagined to be an unmapped area surrounded by country which has either been previously mapped, or is such that control points are easily located in it. Such areas do still exist in the world and it was in relation to one of these that the scheme of experiments here illustrated was started. The errors that would have occurred in this scheme, if each flight had been studied alone without reference to any of the others, have already been discussed and are displayed in Diagram 16. Diagram 17 shows the same scheme

* These marks can conveniently be made on the elastic strings with spots of Chinese White.

after the elastic adjustment has been applied*; on comparing these two diagrams it will be seen at once that the three improvements previously mentioned have to some extent been realised. The calculated positions of the optical centres have, on the whole, moved towards the true positions; thus reducing the average error appreciably. The more serious local errors have been distributed, and the regions of excessive error have been indicated by the excessive distortion of the lines which, before adjustment, were straight.

In this scheme the lateral spacing of the flights was about 20 miles, whereas in an actual survey the spacing would generally be not greater than 10 miles. This was because we realised that there was no chance of our accomplishing sufficient flights to provide a 10 mile spacing, and we preferred to sacrifice the spacing rather than the size of the area covered; for it was one of our objects to show that navigational accuracy can be maintained upon long flights. It will easily be realised that the method of elastic adjustment upon intersecting flights becomes more and more effective and necessary the greater the ratio between length of flight and lateral spacing, hence the method will neither appear so successful, nor so necessary, in the scheme here illustrated, as it would with closer spacing. It was applied to this scheme more with the object of studying the technique of the process, than with any idea that much would be learned from the quantitative results, but in spite of this the results are of interest, and are worth some consideration.

The reduction in the errors brought about by the elastic adjustment in the case under discussion can be seen from the following table, where the errors are given in miles, and refer to the whole of the data included in Diagrams 16 and 17.

---

* In these experiments the numbers of the photographs were, for economic reasons, insufficient to cause them to overlap fore-and-aft, hence it was not in general possible to identify the intersection points of crossing strips from data in the photographs themselves. The procedure adopted for calculating the times of crossing the paths of other flights was as follows. In Diagram 16 the true positions of the optical centres had been plotted by direct comparison with existing maps and, by drawing lines through these points, a good estimate of the true position of any intersection point could be obtained. It was assumed that the instant at which the aeroplane passed the intersection point would divide the time between the exposures immediately before and after this instant, in the same proportion as the intersection point divided the distance between the corresponding optical centres. This procedure would not, of course, be possible in a real survey, but it is closely equivalent to what would be done in a real survey, provided that the intersection point could be identified in both sets of photographs. The difference in procedure can have no appreciable effect upon the conclusions drawn from the experiments and is merely noted here for the sake of completeness. It was, of course, realised that information obtained in England upon the possibility of identifying a single point in two photographs would have no value in other countries, hence it was not held to be expedient to incur the expense of taking sufficient photographs to allow the intersection points to be directly ascertained.

	Without adjustment	With adjustment
50 per cent. error	0·32	0·21
90 per cent. error	0·75	0·40
Maximum error	1·10	0·72

These results may be roughly summarised by the statement that the errors have been reduced by the adjustment in the ratio of two to three.

The reduction of the errors in the distances between successive strips is best seen by reference to the distance between strips E and F where they are crossed by B. The error here is 3·5 per cent. in the unadjusted scheme and 1·07 per cent. in the adjusted scheme, a very noticeable improvement. If we apply this same test to the worst error of all, that between E and F, about 6 miles North of B, we find an error of 5·1 per cent. in Diagram 16 reduced by the adjustment to 3·6 per cent. in Diagram 17. The improvement is not so marked as before, but it is to be noted that, had there been an East-West strip 10 miles North of B, the reduction in this error would have been very much greater. It should also be observed that, had there been an intermediate strip between E and F, the large error in F would have led to a percentage error in the distance between F and the intermediate strip of some 7 per cent., even had the latter strip itself been straight; this error would obviously have been greatly reduced by the adjustment.

Finally let us see to what extent the elastic adjustment would have indicated the existence of the larger errors, had the country been previously unmapped. In this case the only available data would have consisted of the lines of Diagram 17; the true positions of the centres, indicated by dots in this diagram, would of course not have been known. If a straight edge is laid between the control points of each flight in turn, two very pronounced kinks in the adjusted lines will at once be apparent; one in F, where it crosses B, and the other in A, where it crosses C; it will also be observed that there is a pronounced inflection in B where it crosses C and D. The first of these indicates the large error in F that has been fully discussed, the second a serious change of speed that can be seen in Diagram 16 to have occurred towards the end of C; the inflection in B corresponds to a pronounced inflection in the actual path of the aeroplane, due to the gyro rudder control, which was not in good order, precessing until it reached the end of its travel. The three principal errors in the scheme would therefore have been indicated, and their nature suspected. It is again necessary to remark that, had the spacing of the flights been closer, the precise nature of these errors would have been even more clearly demonstrated.

The elastic adjustment has therefore proved to be of distinct value in connection with this scheme, even although the average lateral spacing between flights is little less than half the length of the majority of the flights. We may infer, from these results, that it will be advantageous to arrange for intersecting flights, and to apply some system of adjustment, whenever the average length of the flights between control points is more than twice the lateral spacing, and that it will be *necessary* to adopt some such procedure whenever this ratio is greater than, say, three or four.

The method which we have been discussing is based on the supposition that the ends of each flight spanning unknown country are known in position. Now we have already seen, in earlier chapters, that it is not possible, in the present state of the art of aerial navigation, to arrange that a long straight flight of the nature required in this work shall pass over two widely separated *predetermined* points. It can be made to start over a given point, but it could not, in general, be made to pass over a second given point, unless a deliberate turn were made which would ruin the flight for the purpose in view. It follows that in order to carry out this method successfully one of three alternatives must be adopted; either the country on which the flight ends must be well mapped in detail, so that some known point is bound to occur near the line of the flight, or the second control point in each strip must be fixed *after* the flight has been made, or the aeroplane must carry oblique cameras by which its position can be fixed in relation to control points that lie some distance on either side of the track.

The first alternative would naturally be adopted if the desired detailed survey existed, as it may well do in the case where the flights end upon the sea coast, or upon navigable rivers. The second alternative indicates the advantage of laying down the ground control wherever possible *after* the aerial work has been finished. The third alternative will be considered later.

We have, for purposes of illustration, concentrated attention in the last few pages upon a relatively simple case, in which a large unknown area is surrounded by country that either has been mapped or can be mapped by independent methods. Such cases do occur in practice, but in general it may be said that every separate survey will be carried out under different conditions as regards co-operation from ground surveying, and in each case the methods employed will have to be suited to the problem in hand. At one end of the scale of difficulty comes the problem we have been considering, relating to an unknown area surrounded by country in which the ends of the flights are easily located; at the other comes the survey of areas that are far too large to be spanned in a single flight and in which the control consists merely of a primary triangulation made without reference to the needs of the aerial survey.

Where no previous ground survey exists, everything points to the advantage of carrying out the ground survey *after* the aerial survey; not necessarily after the whole aerial survey has been completed, but after it has proceeded far enough, in any region, to enable the director to see where each control point could be placed to the best advantage. Thus the ground might first be covered with a net of photographic strips spaced say ten miles apart and of any convenient length, and suitable points, near where the flights intersect each other, might be subsequently located by ground methods; the number of intersection points to be located depending of course upon the accuracy required. This ground control might be carried out by triangulation, but it is worth noting that it might be done by astronomical methods, by observers on the ground, carried to the spot, if necessary, by aeroplane. The last method would be particularly useful because the spots located could be chosen without regard to their visibility from other ground stations, and hence would be more easily fitted to the previous aerial net. In country where triangulation is particularly difficult this may be the only practicable way of providing control. We are informed that it is now possible to locate position in this way with an accuracy that is sufficient for the purpose.

We have not done much experimental work upon the problem of fitting a net of navigationally controlled flights to an arbitrary primary triangulation, but we believe that this could be done without very great difficulty. We shall see presently that it is possible to fix the position of the aeroplane in relation to control points situated some miles to one side of it, provided that these points are visible in oblique photographs exposed simultaneously with the vertical photographs, by means of which the navigationally controlled net is primarily constructed.

We have already mentioned our opinion that oblique cameras should always be carried and used during a navigationally controlled series of flights, such as we have had under consideration. We shall now discuss the extent to which such obliques might be used to provide information concerning the ground not included in the verticals, and thus to allow a moderately accurate reconnaissance map of the whole area to be made, without going to the labour of covering the whole ground with vertical photographs.

If the navigational flights are, as we have suggested, spaced some ten miles apart, the obliques, in order to cover the whole intervening ground, will have to be inclined at a considerable angle to the vertical, even when taken from the assumed height of 10,000 ft. We shall see, too, that there are great advantages to be obtained by including the horizon in the obliques, and this again imposes a limit to the amount by which the optical axis can be depressed below the horizontal plane.

Obliques of this nature, when compared with verticals, have several defects, from the point of view of map making. They may not record narrow roads or channels bordered by high trees, nor will they show the opposite sides of steep hills. They cannot therefore give the complete detailed information that is obtainable from verticals; nevertheless the information they can give may be of great value and sufficient for many purposes, especially when we take into consideration that, by their use, it should be possible to produce maps with from one-fifth to one-tenth as much flying per unit area as is required for mosaic work. It must be remembered, in this connection, that when obliques are used in conjunction with the formation of a navigational net of the type described, the ground within each square will be overlooked by numerous obliques, taken from all points of the compass, so that every side of a mountain will come into view, and even narrow channels should be seen end on in some photographs.

The use of very oblique photographs may be limited by haze, which is present on cloudless days in some countries such as England and prevents detail from being recorded at any great distance from the camera. This will probably make the procedure which we are about to discuss unsuitable for use in England or similar countries, but there are many countries in the world in which the air is much clearer than in England, and it is in these only that the very oblique photograph taken from high altitudes will have a general value. It is true that we have in illustration facing p. 90 an oblique taken from some 10,000 ft. in England, in which detail up to 15 or 16 miles can be distinguished, but this, although not as clear as some days that do occur, was yet an exceptional day for England, and it is not possible to adopt, as a routine, a process that can only be carried out on exceptional days.

It is of interest to examine the amount of detail given in the oblique illustrated, realising however that distant objects are in this case much more obscured by haze than they would be in many of the countries in which mapping by the processes under consideration might be desirable. This photograph was taken from a height of about 10,000 ft., from a point over the town of Great Dunmow, in map square E. 3, on the Ordnance Coloured 1 inch to the mile map, sheet No. 98, facing about E.N.E. true. The co-ordinates in this map square, reading from the S.W. corner were 1·12" East and 0·50" North. The square can be drawn by ruling from marks on the border of the map. A 4" × 5" plate with a 6" focal length lens was used and the optical axis was depressed below the horizon by about 17 degrees. The graticule superimposed on this photo-

Oblique photograph taken from 10,000 feet

Oblique photograph with graticule

graph in the lower half of p. 90 will be considered later. For the present purpose it is sufficient to state that the bottom edge of the photograph is about 3 miles from the plumb point, one-quarter the width from the bottom is about 4 miles away, half-way up the photograph is about 7 miles away, whilst one-quarter of the width from the top is about 15 miles away. The horizon line, corresponding to infinite distance, is near the top of the photograph.

From a study of this photograph we can arrive at the following conclusions, which apply in strictness only to the particular country concerned, but which probably have a fairly general application. Woods, fields, hedges and trees of all kinds show up well, being clearly distinguishable up to 10 miles and faintly distinguishable to greater distances of 15 miles and more. Beyond this distance they are obscured by haze and, although in clearer air large woods could no doubt be identified up to very much greater distances, it is clear that they would be so foreshortened as to give little detailed information.

Roads and lanes between hedges are easily distinguished up to about 5 miles away and, in some cases, when they lie in a line with the camera, can be identified up to 8 or 9 miles. No indication whatever is given of roads beyond this distance.

The town of Braintree can be made out, in general outline but not as to detail, at a distance of about 10 miles. It lies just beneath the wing.

In the scheme we are discussing, wherein the obliques are to be used in conjunction with a navigational net of some 10 mile mesh, squares of country 10 miles to the side would be observed from a large number of obliques, taken from points all round the periphery of the square, and it would seem, from a study of this photograph, that few features of interest would be missed, with the exception of small clearings of diameter less than some three times the height of the trees, or say 100 yards; these, if they occurred near the middles of the squares, might not be noticed in any of the photographs. Very large natural features, such as areas of water or wood, will probably be visible at great distances although none appear definitely in the illustration; although these may be too foreshortened to allow their shape to be ascertained, they may have very great value in connection with the determination of the azimuth orientation of the cameras.

In the foreground of this photograph, just under 3 miles distant from the plumb point, practically all important details, down to individual houses, are visible. If the map were to be constructed from parallel flights spaced only 5 miles apart, no point would lie farther than $2\frac{1}{2}$ miles from the track of some flight, so that, with this spacing, a highly detailed map could be made from obliques.

The oblique illustrated does not include ground nearer than some 3 miles from the plumb point, even although the horizon is inconveniently near the top of the photograph. The "vertical" associated with it does not reach beyond about three-quarters of a mile from the plumb point, so that there was a gap left unphotographed some two and a quarter miles wide. If the aeroplane had been equipped with a vertical camera and two such oblique cameras pointing in opposite directions, at right angles to the direction of motion, two strips of country of this width would have been omitted altogether; but if the oblique cameras had been placed with their shorter plate axis horizontal, the gaps could have been reduced to some one and a half miles. Now the field of view of modern aerial cameras, using their shortest focal length, is about the same as that corresponding to the longer axis of these photographs, so that, using these cameras, there would be a gap of a mile and a half between the vertical and the obliques, provided, of course, that only three cameras were used and the horizon included in both obliques.

It would be intolerable to waste the opportunity afforded by the flight of taking this important strip of country, but at the same time it would, as we shall see, be the greatest mistake to miss the information afforded by the inclusion of the horizon and some very distant objects in the obliques. To employ five cameras, however, would seriously add to the expense and trouble of the work. This difficulty could be overcome in either of two ways. Cameras might be provided with an angular field of slightly over 60 degrees, so that three would cover the whole range from horizon to horizon; there would appear to be no fundamental difficulty in doing this, especially as very great accuracy might not be required for this work and aberration could, as we shall see, easily be allowed for in the analysis. Alternatively the obliques, instead of facing at right angles to the direction of motion, could be pointed either backwards or forwards, so that whilst still including the horizon they would join up with previous verticals of the same flight.

It will probably be advantageous in any case to incline the oblique cameras either backwards or forwards, in order to prevent their field of view from being obstructed by the wings, as in the illustration. The exact inclination chosen will depend upon the cameras available and upon convenience of mounting and other considerations, but the larger the field of view available in the camera the less will these angles be restricted by the necessity of including the whole surface of the ground in a succession of photographs taken by three cameras, and the larger will be the area photographed in great detail. With existing aerial cameras a complete cover of the ground could be obtained if the oblique cameras were pointing backwards or forwards by slightly more than 45 degrees.

If new cameras are in contemplation a good compromise might be obtained by using a square photograph of side equal to the focal length and inclining the oblique cameras backwards at 45 degrees, so that the projections of the optical axes of the two cameras upon the horizontal plane would be at right angles to each other.

We thus see that it is possible, by using three cameras simultaneously, to provide at least pictorial information concerning the whole country covered by a series of parallel flights spaced 10 miles apart, and to arrange that the horizon shall be included in all the oblique photographs. We have now to consider the accuracy with which features that are distinguishable in these photographs could be located in relation to whatever ground control is available.

The accuracy with which the aeroplane can be located at any moment in a scheme of this kind has been studied earlier in the chapter, and it is the primary function of the photographs, both oblique and vertical, to enable the positions of the ground features that appear in them to be fixed in relation to these assumed positions of the aeroplane. To some extent it may be possible to readjust our ideas of the position of the aeroplane from an examination of the consistency of the information in overlapping obliques, but we have no evidence as to the increase, if any, in final accuracy that can be effected in this way, and it will be safer to assume that the position of the aeroplane is determined entirely by the navigation process, keeping in reserve any extra accuracy that it may be possible to obtain in a final adjustment.

When the ground is known to be flat, there are many very rapid ways by which the information contained in an oblique photograph can be used to construct a map—as for instance the use of a graticule representing squares on the ground—but the scheme we are considering may be required for hilly country, for which the most obvious and satisfactory procedure would be analogous to that employed when using the "plane table" in ground surveying. Thus the photographs could be used to give the bearing of any point, as seen from the aeroplane at the moment of exposure, and corresponding rays could be drawn on the map from the assumed positions of the aeroplane, as many rays being associated with each point as there are photographs in which that point appears. The mean intersection of all these rays would then be taken as the position of the point in question. This process has the advantage that the location of a point in plan is independent both of the height of the point and of the aeroplane. Finally, when a point has been located in this way its height can be determined, provided that we know the height of the aeroplane and the inclination to the vertical of the line joining it to the aeroplane.

This is the method which we consider should be used for the construction of a map from the photographs taken on a series of navigational flights, and we propose now to consider the errors to which it is subject and the details of its application. The errors may be divided into two groups; in the first group come those due to inaccurate knowledge of the position of the aeroplane and the orientation of the camera, and in the second group are those errors that arise during the identification of ground features and the measurement of angles on the photographs. We shall deal with these two groups in the above order.

The fixing of the aeroplane in plan has been thoroughly discussed and we need only recall here that the "local" accuracies with which the position of the plumb point can be located should be greatly increased when tilt is accurately known, so that the plumb point can be distinguished from the optical centre of the supposed vertical photograph. It is to be presumed that the three cameras would be connected together in some way such that their relative orientation is known and, therefore, if the horizon appears in any oblique, the orientation of the vertical camera can be determined with an accuracy which depends merely on the accuracy with which the horizon line can be located on the oblique. Now we have seen in chapter III that there is evidence that the visible horizon from an aeroplane at 10,000 ft. is to be relied upon to give the horizontal plane with a 50 per cent. error of about 10 minutes. To be on the safe side let us call this 0·2 degree; this is an accuracy five times as great as that which we are led, in chapter IV, to expect from pure skill in flying.

Any error in the estimation of the horizontal plane—that is in the tilt of the cameras—will throw out the position of the plumb point from which the rays leading to the various features to be mapped radiate, and will introduce errors in the plan location of these features of the same order of magnitude as the displacement of the plumb point itself. The local errors of the whole map will therefore be at least as large as the local errors of the plumb points, and we have seen that there is reason to suppose that these can be reduced by the inclusion of the horizon in the obliques, hence it follows that whenever it is important to keep the local errors down, it is important to arrange for the inclusion of the horizon in the obliques.

In the method we are discussing we require to know the bearing, or azimuth, of various features on the ground as viewed from the plumb point, and before we can find this bearing we must in some way ascertain the azimuth orientation of the group of cameras. An error of, say, one degree in the assumed orientation of the camera will result in a displacement of the rays by a distance proportional to the distance from the plumb point, and equal to about 0·09 mile at 5 miles range; that is to say near the centres of the squares formed by our imaginary net. Each point would, however, be defined by the intersections of a number of such rays, so that the actual error in the final location of any point in relation to the plumb points would be somewhat less than the displacement of each ray. Now we have seen that, with good flying, the 50 per cent. deviation of the azimuth of the camera from the mean orientation during a flight should be in the neighbourhood of one degree, so that, by very careful attention to the "swinging" of the compass, it should be possible to fix the azimuth orientation of the camera at the moment of any exposure with this accuracy   The errors of location of points within the squares relative to the plumb points could thus be kept within reasonable bounds, although they might be sufficiently large to be distinctly undesirable.

In general, however, it should not be necessary to rely upon fixing the azimuth of the camera by direct reference to the compass, for should it happen that either of the obliques contains a single ground feature which can be identified in the verticals of one of the adjacent flights, this common point could be used to determine the orientation of the camera group, with an accuracy depending merely upon the errors of the navigational scheme itself. Where this can be done, therefore, it will be clear that the only additional inaccuracies introduced during the extension of the map from the plumb points to the regions between the flights will be due to inaccuracies in the measurement of angles within the photographs themselves.

Now at 10 miles range the width of country covered by a wide angle photograph is over 8 miles, whilst from 10,000 ft. the width of the vertical photograph is about $1\frac{1}{2}$ miles, so that there will, in general, be an area of ground of about 12 square miles common to each oblique and to the verticals of the adjacent flight. There are two obliques to each group of cameras, so that there will be 24 square miles of country from which to choose such a point. A study of the oblique facing p. 90 in the area for which the depression below the horizon lies between 10° and 11·5° will show that, except in quite featureless country, there is a very good chance of such a common point being identified. For this purpose of fixing the azimuth orientation of the cameras, very distant objects, visible from 30 miles or more, will be of great value; one conspicuous object, such as a mountain top or large lake, might serve to fix the azimuth of very many photographs.

We have now reached the important conclusion that it should, in general, be possible to fix the azimuth orientation of each group of photographs through the medium of ground features common to the obliques of the group and the verticals of other flights, and that, when this can be done, the map may be extended from the plumb points to cover the ground between the flights without serious loss of accuracy, other than that resulting from inability to measure accurately the angles subtended between points that occur in the same group of photographs. We shall turn therefore to consider the second group of errors, namely those errors introduced by imperfect identification of ground features, and by the inaccurate measurement of angles in the photographs.

The choice of a suitable method of measuring the relative azimuth angles in the obliques calls for some consideration. To be in keeping with the requirements of the scheme under discussion, any method which is used for this purpose must be very rapid, because the survey of very large areas by these methods will involve the analysis of large numbers of photographs, and any great time spent on each photograph will add seriously to the cost of the survey. On the other hand great accuracy is not required, a tenth of a degree being ample. These conditions clearly indicate the use of some form of graticule, which, for the work in hand, would naturally take the form of a series of lines, each of which represents all points having the same azimuth in relation to the plumb point. The form taken by these lines will depend upon the tilt of the optical axis of the camera when the photograph was taken. A separate graticule will therefore be required in theory for every tilt. The construction of these graticules is discussed in Appendix IV, where it is shown that, for accuracies of a tenth of a degree, it is sufficient to provide one graticule for each integral degree of tilt, and to use the one that is most nearly suited to the actual tilt of the photograph.

A graticule for this purpose consists of a straight line representing the horizon, intersected by a pencil of straight lines diverging from the plumb point which, in the illustration, is situated outside the limits of the photograph. One of these lines, which may be designated zero, is perpendicular to the horizon line and must pass through the optical centre. The graticule is adjusted to the photograph by placing the horizon line in the graticule upon the horizon line in the photograph, in such a way that the zero azimuth line passes through the optical centre of the photograph. The correct graticule to be used with any given photograph must be chosen after measuring the distance between the optical centre and the horizon line of the photograph, for this distance is proportional to the tangent of the angle of depression of the camera.

It has been shown that with good flying the unintentional tilt can be kept within two degrees and that it is sufficient to provide one graticule for each integral degree of depression of the optical axis; hence a series of seven graticules, covering a range of three degrees on either side of the prearranged depression, should be amply sufficient to meet all requirements of a particular survey. If these graticules are conspicuously labelled and arranged in a suitable rack, very little time need be expended in choosing the correct one for use with any given photograph.

It will of course be realised that the true horizon on the photograph will be above the visible horizon, by an amount that depends upon the height of the aeroplane above the haze, or above the ground horizon when there is no haze. It is sufficiently accurate for the purpose to assume that this dip of the horizon, expressed in minutes, is equal to the square root of the height of the aeroplane above the horizon, expressed in feet. The height of the top of the haze would of course be recorded when ascending to and descending from the survey height.

The oblique photograph illustrated was taken at the same moment as a vertical photograph, the two cameras being fixed together in such a way that the optical axis of the "vertical" was truly vertical when the longer side of the oblique photograph was horizontal and its optical axis depressed by 16 degrees below the horizontal plane. The horizon could be traced on the oblique photograph, so that it was possible to determine the inclination of the axis of the vertical camera, and hence the position of the plumb point, with an accuracy that depended only on the accuracy with which the horizontal plane could be deduced from the visible horizon. This plumb point could therefore be identified on existing maps and the azimuth orientation of the cameras could then be determined by reference to any one other point that is visible both in the map and the oblique. The azimuth angle between this point and any other point could then be found both from the map and photograph, and the results compared to give an indication of accuracy with which the graticule can be used to measure relative azimuth angles of points in the same photograph.

This has been done for the oblique illustrated and one other, and the errors are recorded in Table 5, both in the form of degrees and of the distance between the point and the ray that should have passed through it. It will be seen that, for the most part, the errors are small compared with the errors involved in the navigational process, so that we may draw the conclusion that it is possible, by the method used, to identify and estimate with rapidity the relative bearing of points in a given photograph, and that with an accuracy quite sufficient for the purpose in hand.

In both these photographs the true horizon was taken to be 0·75 degree above the visible horizon and the depression of the optical axis was then found to be 17·2 degrees in each case; the graticule was made for a depression of 17 degrees. The visible horizon was faintly, but definitely outlined on the negative, but cannot be so easily distinguished in the reproduction.

The curved lines in the graticule, which intersect the azimuth lines approximately at right angles, correspond to points of equal depression

Table 5. *Errors in the analysis of two oblique photographs by the graticule method.*

Photograph A (facing p. 90).

Positions of test points		Azimuth errors		Depression errors	
Azimuth	Distance from Plumb Point	Degrees	Miles	Degrees	Feet
+ 19·0	2·92	+ 0·3	0·02	+ 0·3	90
− 3·3	2·75	+ 0·3	0·02	+ 0·3	80
− 19·2	3·11	− 0·2	0·01	0·0	10
+ 20·5	5·46	+ 0·4	0·04	+ 0·3	160
− 0·8	4·01	0·0	0·00	+ 0·3	140
− 15·7	4·30	+ 0·4	0·03	0·0	10
+ 16·9	8·82	+ 0·2	0·03	+ 0·2	190
+ 5·1	8·92	+ 0·1	0·02	+ 0·2	180
− 15·9	9·20	− 0·3	0·05	+ 0·1	50
+ 10·0	11·23	− 0·1	0·02	+ 0·2	220
− 9·5	12·74	− 0·2	0·05	+ 0·1	80

Photograph B (not illustrated).

+ 19·0	2·99	0·0	0·00	+ 0·3	90
+ 19·0	3·64	0·0	0·00	+ 0·4	150
+ 1·9	2·91	− 0·1	0·01	+ 0·2	70
− 13·9	3·00	0·0	0·00	+ 0·1	40
+ 19·2	8·06	+ 0·1	0·02	+ 0·4	340
+ 2·8	6·42	0·0	0·00	+ 0·2	140
+ 9·0	8·06	0·0	0·00	+ 0·3	230
− 6·1	7·80	+ 0·1	0·02	+ 0·3	210
− 5·9	4·64	+ 0·1	0·01	+ 0·2	100
− 18·2	8·50	+ 0·2	0·03	+ 0·2	160
+ 13·8	11·55	+ 0·1	0·02	+ 0·3	380
− 12·8	12·40	+ 0·1	0·02	+ 0·3	330
+ 7·5	16·70	0·0	0·00	+ 0·4	470

The point used to find the azimuth of the camera was in each case near the zero azimuth line and at about 10 degrees depression from the horizon, or about 10 miles distant from the plumb point.

below the horizon and, like the azimuth lines, occur at intervals of one degree. A simple method of constructing these lines is discussed in Appendix IV. It is convenient to be able to read off the depression of any point below the horizon for three reasons: in hilly country it is in this way possible to obtain some idea of the height of the ground, in flat country a knowledge of the depression of a point will help to fix its position relative to the aeroplane, and finally, if the height and position of some point are independently known, a knowledge of the azimuth and depression of the point and of the height of the aeroplane will allow the position of the aeroplane to be fixed relative to this point The last may, as we have seen, be a matter of great importance in connection with the problem of fitting navigationally controlled net to an existing ground triangulation, in which the control points may not always lie close to the line of any one of the flights.

In order to determine the accuracy with which the depression of a point can be estimated, the depressions of chosen points in the above two obliques have been determined, using the equi-depression lines in the graticule. The distances of these points from the plumb point could be found from the ordnance map, and the height of the aeroplane could be calculated by comparing the scale of the vertical photograph with that of the map. The true depressions of the points could therefore be determined and compared with the depressions estimated by the graticule. This has been done and the errors, or differences between these two independent methods of determining depression, have been recorded in the fifth column of Table 5. It will be noticed that, in each case, there is a consistent, or mean, error of about 0·2 degree, which suggests that the estimated position of the true horizon in both photographs was incorrect by this amount. The 50 per cent. deviation from the mean error is, in both photographs, no greater than 0·1 degree. This represents the random errors introduced during the whole process of identifying a point on the map and photograph and estimating its depression below the assumed horizon.

The last column in this table shows the errors which would have been made had the heights of these points been estimated from the depressions found in the oblique photographs. It will of course be realised that these errors do not represent the whole of the errors which would occur in the estimation of height during a survey of the kind contemplated, for the assumed height of the aeroplane, and the assumed distance of the point from the plumb point, would themselves be subject to errors, which would influence the estimate of height. The figures given in this table do no more, therefore, than indicate the order of error which may be introduced during the rapid analysis of the photographs and the identification of natural objects on the film; they show also that in one particular instance the horizontal plane was estimated from the visible horizon with an error of 0·2 degree, which happens to be very close to the 50 per cent. error that was found in a series of experiments by the Air Ministry upon the determination, by this means, of the horizontal plane (see chapter III). We must, however, rely on the Air Ministry experiments for cumulative evidence relating to this last point.

A short discussion of the accuracy with which height might be determined by these methods, in an actual survey, occurs in Appendix V, where the conclusion is reached that the accuracy, in the absence of any horizon trace on the photograph, will be very low, but that, if the horizon can be included, the 50 per cent. error of the final height estimation of any point should be of the order of 100 ft. This is, of course, the error in determining absolute height above sea-level. Local slopes and hill formations can be very effectively seen by viewing overlapping photographs simultaneously in a common stereoscope. Considerable success in sketching ground formations by this means was achieved in Palestine during the war and in Egypt after the war. The authors have not had time to study this process in any detail, but a specimen stereoscopic pair of oblique photographs, taken in Palestine, from a height of about 4000 ft., will be found in a pocket in the cover; if this is viewed in a common stereoscope, some idea will be obtained of the extent to which the form of the ground can be estimated in this way. These photographs have been selected from a number of stereoscopic pairs supplied by Mr Hamshaw Thomas.

Let us now summarise the arguments advanced in this chapter and collect the experimental results achieved.

It has been shown that an aeroplane can be flown for a long distance in such a way that, if the positions of two points near the beginning and end of the flight are known, the position of the aeroplane at any instant during the flight can be fixed with considerable accuracy, on the simple assumption that the aeroplane passed over the ground, between the fixed points, in a straight line and at a constant speed. It is suggested that this might form the basis of a system of aerial mapping which, although not so detailed as the mosaic method, would be extremely rapid and economical; and which would, unlike the mosaic method, be equally suitable for flat or hilly country. The scheme suggested is to cover the ground by two series of long parallel flights, the traces of which over the ground would form a rectangular net. During these flights the aeroplane would carry three wide angle cameras, which would be so arranged that the whole of the ground eventually appears in photographs, at sufficiently close range for most of the important details to be distinguished and fixed in position. The map might be constructed from these photographs in a manner similar to that used with the "plane table" on the ground, each point being located at the intersection of a number of rays drawn from the plumb points at the moment of exposure; the plumb points themselves would be located by the navigational process previously mentioned. Finally, it is suggested that the absolute height of selected points could be estimated from the depression of the points below the horizon in all the photographs in which they appear, and that steep slopes and local height differences could be sketched in by stereoscopic methods.

The process involves the fixing, by independent means, of at least two points upon each straight flight; and since it is impracticable to arrange for the flights to pass accurately over more than one pre-arranged point, it may be necessary either to fix some of the control points after the flights have been completed, or to be able to locate the aeroplane through the medium of oblique photographs by reference to control points that lie at some distance to one side of its path. It is not, however, essential to provide two *ground* surveyed control points for every straight flight, for some points may serve to locate more than one flight, and some flights may be located by the fact that they cross other flights that have themselves been located. Even in the extreme case where the ground control consists merely of a primary triangulation laid down without reference to the requirements of the aerial work, no insurmountable difficulty is anticipated in adjusting a navigationally controlled net of flights to the control points, provided that the latter can be distinguished in some of the oblique photographs.

It has not been practicable for us to test these ideas, as we did in the case of the mosaic method, by the construction of a complete map, but the various processes involved have each been studied separately, and from the data so obtained an estimate has been made of the errors that may be expected to occur. These errors may be summarised as follows.

The "local" error, that is the displacement in any given direction of a point relative to its immediate surroundings, may be expected to be such that the 50 per cent. error will lie between 0·01 and 0·03 mile; the lower figure applies when the horizon is visible upon the obliques, so that unknown tilts can be practically eliminated, whilst the larger figure assumes that there is no independent check on tilt, other than reliance upon the pilot's skill in flying.

The "local" errors will, of course, be independent of the closeness of the ground control, unless it is close enough to provide control upon each separate photograph; the "total" error of a point relative to the control will, on the other hand, depend directly upon the closeness of the control. For single flights standing alone, the maximum error to be expected in any flight should have a 50 per cent. chance of being less than 1 per cent. of the distance between control points, but when a net of intersecting flights is available it is possible to reduce these errors considerably, by a process of distribution which can be conveniently carried out mechanically. In the experimental net of flights, spaced 20 miles apart, which was laid over an area of about 90 × 50 miles in the Eastern Counties, the 50 per cent. error, after this distribution, of any point relative to the supposed known edges of the net, was 0·2 mile, or some 0·4 per cent. of the width of the supposed unknown area. The maximum error was 0·7 mile or 1·4 per cent. of the width, but this error was traced to a cause that should not recur, when the danger of its occurrence is realised. The 50 per cent. error of the length of any one side of the squares of the net—in this case in the neighbourhood of 20 miles—was 0·7 per cent. of the length of the side.

It was next shown that it should be possible to locate ground features lying between the flights, through the medium of oblique photographs, without serious loss of accuracy; so that the accuracy of the whole map should be of the same order as the accuracy of location of the plumb points indicated in the preceding paragraph.

Finally it was shown that, provided the horizon—haze or real—appears in one of the photographs of each group of three, it should be possible to settle the absolute height of any well-defined point with a 50 per cent. error of about 100 ft., and it was shown that the forms of hills and valleys could be estimated by viewing overlapping photographs in a common stereoscope.

CHAPTER VII

EQUIPMENT AND DETAILED PROCEDURE

IT is now proposed to describe, in rather more detail than has hitherto been convenient, the apparatus that we have employed and the procedure that we have adopted in the operations of which a general account has been given in preceding chapters.

When laying down any procedure that has to be carried through in the air, it is important to realise that processes of thought which may appear easy to a man seated in a comfortable and warm office will be very much more difficult to carry out efficiently when in the air. Naturally everyone will make some allowance for this difference but, in the absence of actual experience, the tendency is to under-estimate it. Surveyors know the difference between thinking out a problem in the office and in the field. Let a surveyor imagine himself to be taking theodolite readings, from an exposed situation in cold and windy weather; let him increase the cold and the wind beyond anything he has ever experienced, add a continuous deafening noise, and let some element of fright and excitement be introduced, for example let him imagine that an infuriated bull is on the other side of a fence which may just conceivably give way at any moment; finally let him imagine that he is slightly unwell, as if recovering from an attack of sea-sickness. He will now be in a position to understand something of the conditions in which the aerial observer is constantly working, and he will realise that these conditions are not exactly favourable to original thought.

This simile may be fanciful, but it is not overdrawn. The slight sickness corresponds to the effects of the rarefied atmosphere in which the observer is working; the continual, though slight, fear of the bull corresponds to the almost subconscious listening for a change of engine note that would pressage a forced landing, perhaps in difficult country; while certain factors, such as the hampering effect of thick clothing and restricted space, have not been represented. Whatever the true cause, it is a matter of experience that observers are liable to make mistakes in the air of a nature which they would never be expected to make on the ground, and they are often at a loss to account for their own apparent stupidity, when they review the circumstances in the comfort of the office.

The moral of this is that any procedure to be followed in the air must involve a minimum of independent thought on the part of the crew. The sequence of operations must be as straightforward as possible, and it should be thoroughly familiar to the crew, or be written out in large type and fastened in a conspicuous place. Calculations must be reduced to the simplest possible slide rule operations, or to the mere identification of points on curves. Mechanical operations, such as the manipulation of the camera, must be simple and so familiar as to be carried out almost subconsciously. There is thus much to be done between the formulation of the theory underlying some operation and the development of the corresponding routine to be used in flight.

With these preliminary remarks, we shall now deal with the various items of the equipment and with the routines involved in their use.

THE AEROPLANE. The most important item of the equipment is the aeroplane itself. We know of no existing aeroplane which is entirely suitable for aerial surveying; the type used in our experiments—the D.H. 9 A, illustrated facing p. 34—though satisfactory in some respects, has certain serious defects. Although we have had no opportunity to experiment with other types of aeroplane, we feel that our experience has been sufficient to enable us to define the principal characteristics that should be possessed by any aeroplane which is to be used for general purposes in aerial surveying.

Most important of all, the aeroplane must be easy and safe to land in difficult places. Whether it should be a seaplane or land plane will depend upon the character of the country over which it will be flying, but in either case it must be designed to give the maximum safety in a forced landing upon the country over which it will be working. These requirements may lead to the use of an amphibian, or possibly to a convertible aeroplane which could be fitted up either as a land plane or as a seaplane, according to the country over which it is working. In any case it will be important to keep the loading per unit area very low, and to design the controls to be as effective as possible at the landing speed: it may be easier to meet these requirements in a surveying aeroplane than in one designed for transport, because a high cruising speed is not so essential in the former as in the latter.

The reason for this insistence upon safety in a forced landing is, of course, that no commercial venture could withstand the excessive losses, both in equipment and personnel, which would result from an undue proportion of accidents. For the same reason it is of vital importance to employ only the most reliable engines obtainable.

Having provided as far as possible for the safety of the crew and equipment, the factor next in order of importance is the pilot's view in a forward and downward direction. We have seen that an essential operation in an aerial survey is the bringing of the aeroplane accurately over a predetermined point, whilst flying on a given bearing; a difficult operation under any circumstances, but one in which the difficulty will be seriously increased if the point is liable to disappear just at the time when the pilot most desires to see it. The illustration facing p. 112 gives some idea of the way in which the pilot's downward and forward view in the D.H. 9 A is obstructed by the lower wing and undercarriage of the aeroplane. The space left, between the wing and the body, is intended to allow the pilot to keep the objective in view as he approaches it, but when there is any cross wind causing the aeroplane to drift sideways, the objective, as seen from the pilot's seat, does not appear to pass down the slot at all, but is obscured by the wing at the very time when it is most important that it should be visible.

The pilot who made the mosaic described in chapter v was, on the whole, very successful in getting over the starting-point of each strip, but he found the operation difficult and he states that he approached the starting-point upon a curve which allowed him to see it through the slot during the whole approach. This procedure is bound to result in the compass being in error at the start of the strip, owing to the setting up of disturbances during the curved approach. The pilot was, however, equipped with the gyro fixed azimuth, so that he could swing his aeroplane quickly on to the calculated bearing, without waiting for the compass to settle down, and it is doubtful whether such good results could have been obtained from a tractor giving so bad a view as the D.H. 9 A, unless the gyro fixed azimuth had been used.

It should not, however, be impossible to design an aeroplane having a very much better view than the D.H. 9 A, and if this were done the operation of getting over the starting-point would be much simplified. The pilot could then manœuvre so as to bring the objective on to the required bearing relative to his aeroplane whilst still at some distance from it, and could approach it whilst flying in a straight line upon the calculated course, making only small corrections as he approaches more and more closely. With such an aeroplane it might be possible to do good work without the gyro; this question is discussed in more detail later.

An alternative procedure would be to allow the observer to direct the pilot over the objective, and when the observer has little else to do, this method has been found very effective, both because it avoids the errors introduced when a pilot looks down too continuously at the ground beneath him, and because it is easier to provide a good view for the observer, whose position in the aeroplane is not so restricted as that of the pilot. In aerial surveying, however, the observer has so much to do that it will probably be better to put the whole responsibility of finding and getting over the starting-points upon the pilot, and if this is done great care should be taken to ensure that the pilot has the best possible view. In cases where an aeroplane with a good pilot's view cannot be obtained, the alternative of letting the observer direct the pilot might be considered, but this procedure is not recommended unless it is unavoidable.

An additional requirement in an aeroplane to be used for surveying by the methods we have described, is that the observer should have a good view, preferably over both sides, in a rearward and downward direction. This is to enable him to take drift bearings to find the wind at the survey height by the method described towards the end of this chapter. In the D.H. 9 A we work over one side only, because the airscrew blast in this machine is much more severe on the port than on the starboard side of the observer's cockpit. It would, however, be much more convenient to be able to work from either side, because, if one side only is used, it is occasionally necessary for the observer, when taking a tail bearing, to watch a point that is drifting under the body of the aeroplane, and this makes him lean far out over the side in an uncomfortable and somewhat dangerous position.

It is of some importance to consider the relative advantages of the pusher and tractor in the light of the above requirements. The pusher has the immense advantages that it is relatively easy to arrange for a good pilot's view, and that neither the pilot nor the observer need be in the airscrew blast. The last point is more important than might be supposed; it is probably the principal reason why the F.E. 2 B was such a favourite aeroplane in experimental squadrons during the war. The disadvantages of the pusher are that it may be more difficult to provide the observer's downward and rearward view, and that the pilot may possibly find himself handicapped by the absence of the wings in front of him; in the tractor these undoubtedly help him to maintain the lateral level of his aeroplane with respect to the horizon. We have no experience upon the latter point, but think that any loss of accuracy might be avoided by providing a special horizontal bar in front of the pilot, to take the place of the wings. It is generally considered to be more difficult to obtain a good performance from a pusher than from a tractor, but it is probable that this difference, if it exists at all, applies more to top speed than to climbing power, and we have already observed that a high cruising speed is not essential. On the whole we are of the opinion that a specially designed surveying aeroplane should be a pusher, but we also consider that it would not be impracticable to design a good tractor for this purpose, should there be any strong objection to employing a pusher.

It is of considerable importance that the aeroplane should be stable at its cruising speed, and that it should be possible to "trim" each of the controls separately. The ideal to be aimed at is that the aeroplane should be able to fly itself, and that it should be equipped with a slow motion adjustment on each control, so that the pilot can trim the aeroplane until, on abandoning the controls, it continues to fly straight, at a constant speed, with the wings level. These requirements are more important than might be supposed; experience shows that the more the aeroplane can be left to itself the better are the results obtained. The D.H. 9 A, in common with most aeroplanes of its class, possesses a slow motion adjustment on the fore-and-aft control, but not upon the lateral or directional controls.

All three controls should be designed in such a way that back-lash in the mechanism by which they are actuated can be avoided. This applies to the throttle lever equally with the other controls; it must not be forgotten that, in this work, the throttle is as much a control as the rudder or elevator and should therefore be made large and capable of very fine adjustment, with no possibility of altering its position unless purposely moved by the pilot.

With regard to performance, it is essential to be able to reach the survey height without undue loss of time, and to be able to fly at that height with a considerable reserve of power. This implies that the "ceiling," or maximum height at which the aeroplane can fly, must be considerably above the survey height. It is important, in this connection, to remember that the ceiling of an aeroplane, which has been in service for some time in tropical countries, will be considerably lower than the official ceiling, as given by tests on a new aeroplane in England. The official ceiling of an aeroplane designed for general surveying purposes should therefore be considerably above the height at which the survey is to be carried out.

The cruising speed should not be less than about 75 miles per hour at 10,000 ft. If it is much less than this, the aeroplane will be too much at the mercy of the wind, and there will be many days upon which the speed, when flying up wind, will be too slow for practical purposes. From this point of view a higher cruising speed would be better, but it may be advisable to be content with a cruising speed of some 75–80 miles an hour, on account of the difficulty of combining a high cruising speed with a low landing speed.

The observer's cockpit should allow plenty of room and afford for him, when sitting down, adequate protection from the wind, but the sides must not be so high that he cannot easily lean over to see where he is and to take "tail bearings." On no account should any reliance be placed upon seeing through holes in the floor, unless it can be arranged that his head comes out of the hole. The small area that can be seen, from a sitting position, through any ordinary glass panel in the floor is almost useless. There is no harm in providing a glass panel, provided that it is not used as an excuse for restricting the observer's view in other respects.

THE CAMERA. Next to the aeroplane, in order of importance comes the camera. We shall not enter into the technical details of the camera from the photographic point of view, because we have not studied this side of the problem; we shall merely consider the camera from the point of view of the operations to be performed upon it, assuming that it is of a type which will give satisfactory photographs. We have worked throughout with the Air Ministry L.B. camera, which uses 5 × 4 inch plates, carried in boxes of eighteen, and which can be fitted with alternative lenses of different focal lengths. For various reasons we do not consider this camera to be suitable for use in extensive surveying schemes, but it was the most suitable of those we could obtain, and it has served our purpose.

In all our work we have used the shortest available focal length—six inches—with the object of obtaining the widest angle of view and thus covering the largest area of ground in each photograph. The choice of a suitable angle for the field of view of the photograph is a matter of great importance; if, in the mosaic work, this angle is too small, the amount of flying required to cover a given area of ground from a given height will be unduly increased and, what is much worse, the difficulty of avoiding gaps between successive strips of photographs will also be increased; if, on the other hand, this angle is too wide, small accidental tilts in photographs which are intended to be vertical will cause serious errors. The optimum angle will depend upon the accuracy of the flying and upon the value placed upon economy in operation, as opposed to accuracy in the final map. For rough mosaic mapping, where great economy is required, a very wide angle lens might be used, if it can be obtained; whereas a smaller angle would be more suitable where accuracy is desirable and economy is not of first importance. In our case we merely used the widest angle lens that we could obtain, and the accuracy of the mosaic described in chapter v indicates that this angle was not so wide as to introduce serious errors. In connection with the rapid methods of mapping described in chapter vi there is, as we have seen, every reason to use the widest angular field obtainable, up to 60 degrees, but there is little object in going to an angle much greater than this.

When, as in the L.B. camera, the plate is not square, the longest dimension should always be placed across the direction of motion, because the amount of flying required to cover a given area, and the difficulty of avoiding gaps, depends upon the width of the strip of photographs; the length of ground covered by the photographs in the fore-and-aft direction merely influences the frequency of the exposures, and does not affect the total distance flown. In a camera specially designed for aerial work there does not, however, appear to be any reason why the photographs should not be square, because if the lens can cover a given angle in one direction, it can presumably do so in any other.

The size of the photograph is of no particular importance from the operational point of view. It is decided merely by the amount of detail which is required in the photographs. Cameras which are being developed by the Air Ministry and other bodies, for aerial surveying, give, as a rule, photographs about 8 × 10 inches square. With the widest angle lens in general use, operating from about 10,000 ft., this will give mosaic maps on a scale of about 6 inches to a mile, and detail less than the size of a horse and cart will be visible. It is quite possible that smaller photographs than these would give all the detail required with greater economy when surveying large areas for relatively small scale maps, but it is possible that the larger size will be convenient for general purposes. If, however, these large sized photographs are used, it is practically essential to work with films, on account of the great weight of plates. It is also easier to cause large numbers of films to be fed continuously than is the case with plates. In mosaic mapping, with its necessarily short flights, this might not be a matter of great importance, but in the type of mapping contemplated in chapter VI, involving long straight flights in which large numbers of photographs must be taken at regular intervals of about half a minute, it is of the greatest importance that the supply should be continuous.

Although the mosaic method of mapping, described in chapter V, does not require the camera to be very accurately adjusted, yet it is advisable always to use a camera in which the internal adjustments are very accurate; thus, the focal length should be accurately known, the lens aberration small, and some provision should be made for indicating upon each photograph the optical centre and at least one datum line. The reasons for this are that some classes of work, for example that described in chapter VI, require this information, and in any case it may be desirable to re-sect the photographs when, at a later date, a sufficient number of control points becomes available.

All cameras designed for aerial surveying should contain devices by which a serial number and the height at the moment of exposure can be automatically recorded upon each plate. In addition it is useful to record the tilt of the camera to the apparent vertical at the moment of exposure, and it might be advantageous to record time down to fifths of a second. Provision for the first three of these records is made in most modern surveying cameras, but not in the L.B. camera.

The films, or plates, of the camera should be changed mechanically, and the exposures made automatically at regular intervals adjustable by the observer; it should also be possible to start the first of a series of photographs at any given moment. It is, however, of the highest importance to arrange that the automatic gear can be easily thrown out of action, and the camera actuated entirely by hand. The operating levers, and in fact all the operating parts, should be made large and easily handled by a man with cold hands encased in thick gloves. This is a remark that should be applied to all instruments that are to be used in the air; it is a point in design that is often overlooked.

THE CAMERA MOUNT. We have now to consider the mounting of the camera in the aeroplane. Great care should be taken to see that the camera and all its adjustable parts are easily accessible to the observer, who may have to manipulate them for three hours or more at a time, under conditions of considerable discomfort. The next point to be considered is that the camera must be guarded against vibration, but at the same time held positively in a given position relative to the aeroplane. This difficulty has been successfully overcome by the Air Ministry, and other makers of aerial surveying cameras, and it will not be further considered here.

For reasons discussed in chapters IV and V, it is necessary* to arrange that the whole camera or group of cameras can be tilted through 4 or 5 degrees in any direction in order to neutralise accidental tilts in the aeroplane, and that it can be rotated through some 40 degrees about a "vertical" axis, to allow for the drift of the aeroplane in a cross wind. The standard mount for the L.B. camera, as we received it from the Air Ministry, was adjustable only for fore-and-aft tilt, so that the other two motions had to be added, after our experience had shown them to be necessary. The converted mount is shown facing p. 112; it is, of course, more cumbersome than would be necessary if it were re-designed from the beginning. The lowest frame in this illustration is attached rigidly to the aeroplane, and above this is a square frame which provides the lateral adjustment by rocking upon the lower frame; this motion is operated through the non-return screw with disc head on the extreme left of the illustration. A third frame rocks upon the second frame, and is controlled by a second disc-headed screw which is not visible in the illustration. The fourth frame, or platform, carries the U-shaped angle iron camera

* This may not be absolutely necessary in the methods of chapter VI provided that it is certain that the horizon will be visible on the obliques.

bearers, and is suspended from the third frame by means of bell cranks which are controlled by the external spring and lever visible on the left-hand side of the third frame. The camera and its bearers can be rotated upon the fourth frame and clamped in position by means of two wing nuts working in circular slots, one of which is visible in the illustration. The edge of the brass disc in which these slots are cut is graduated to show the number of degrees through which the camera has been rotated. The camera itself is supported through the medium of pieces of "Sorbo" rubber sponge, to absorb short period vibrations; it is adjustable relative to the carriers on a three point system, which allows the axis to be alligned truly parallel to the axis of rotation.

The camera carries two spirit levels at right angles to each other, clearly visible in the right-hand top corner of the illustration. It will be remembered that the reason for providing adjustments upon the tilt of the camera axis relative to the aeroplane is to allow this axis to be adjusted parallel to the apparent vertical, so as to neutralise any consistent tilt in the aeroplane. Spirit levels, of course, indicate the apparent vertical. The procedure we have adopted, when making a photographic flight, is to alter the tilt adjustments only when the spirit levels show a consistent error over some considerable period of time, and to ignore entirely minor fluctuations enduring for less than, say, one minute. In a procedure of this nature slow motion screw adjustments, of the type provided, are satisfactory and have the advantage that, being of a non-return nature, they do not demand continual attention. An alternative device, which has been used in America, is to mount the camera upon gimbals and connect it by a link motion to a control stick similar to the control stick of the aeroplane. This control stick moves parallel to the axis of the camera, and carries upon its top a spherical spirit level. The observer holds the control stick all the time and adjusts it so as to bring the bubble of the spirit level central at each exposure. We have not tried this device, and are therefore unable to compare its action in practice with the device that we have used. All that we can say is that the latter has been satisfactory.

The methods of chapter VI would require a triple camera mount upon which the vertical and oblique cameras could be mounted together. The design of this mount and the fitting of it into an aeroplane, so as to allow for adjustment and yet obtain the necessary field of view for all the cameras, will present some nice problems which we have not studied; they are problems that will be much simplified when the aeroplane is specially designed for the purpose, but they should not prove insuperable in any case.

GYROSCOPICAL INSTRUMENTS. Although the aeroplane and the camera are the two principal items in the equipment of an aerial survey, there are several other instruments which must be carefully chosen. The two gyroscopical instruments which have been mentioned will not here be described in detail; the specimens we have used were developed by and are the property of the Air Ministry, and it will be necessary to make application to that body if it is desired to use them.

The essential feature of both instruments is a small freely pivoted gyroscopic wheel, which is so accurately balanced that it will maintain its orientation practically unchanged for many minutes, independently of the movements of the aeroplane. In the "gyro azimuth indicator" this wheel is attached to a pointer which will indicate changes of direction of the aeroplane and so act, for a time, as a supplementary compass. In the "gyro rudder control" the wheel actuates the rudder through a delicate pneumatic relay and so keeps the aeroplane automatically upon a straight course. When accurately adjusted, the latter instrument will hold the aeroplane accurately on a given course for a very long time, but its main function in surveying is to relieve the pilot from the necessity of keeping his feet on the rudder bar and his attention fixed upon the delicate operation of keeping straight; any slow rate of turn, due to precession of the gyro wheel, can be corrected from time to time by the use of a small control lever, which adjusts the relationship of the gyro wheel to the rudder.

To be of value in a survey, these instruments should be able to maintain their orientation within one degree for at least 10 minutes. The "gyro rudder control" should possess a slow motion adjustment whereby the course of the aeroplane can be altered relatively to the gyro wheel, through at least 20 degrees on either side of a central position. When the gyro rudder control is in action, the compass will be very steady in the steady air normally found at the surveying heights, and the pilot can quickly detect any permanent change of course, due to precession in the gyro, and correct it by means of this adjustment. If no such adjustment is provided, the aeroplane could not be brought back upon the desired course, except by breaking the connection between the gyro and the rudder, centralising the gryo and starting afresh; in our experience this always involves a temporary variation in the direction of flight, such as that observed at about the 13th photograph of flight *B* in Diagram 16.

THE COMPASS. Another instrument which must be very carefully chosen is the compass. We have already discussed in chapter III the reasons why it is difficult to obtain accurate results from a compass carried in an aeroplane; these difficulties vary enormously with the type of compass used. A compass to be used for aerial surveying, besides

Showing pilot's view on D.H. 9 A

The L.B. camera on mount

The Aperiodic Centesimal compass

The strut thermometer

having the necessary characteristics discussed in chapter III, must be provided with a very open scale, which can be read with ease by the pilot within one degree. The only compass we know which combines these qualities is that known as the Aperiodic Centesimal. It is illustrated facing p. 112, where it will be seen that the conventional card is replaced by four wires bearing the numbers 0, 1, 2 and 3. The course upon which the aeroplane is flying is found by adding the number on the scale that lies opposite the end of the wire to the number on the wire multiplied by 100; thus, the compass illustrated is indicating a course of about 294 degrees. When mounted in an aeroplane flying horizontally, the ends of the wire pointers come much nearer to the scale than in illustration, in which the compass was tilted for photographic convenience.

THE ALTIMETER. An altimeter is required, as we saw in chapter III, for two distinct purposes; to enable the pilot to maintain a constant height, and to provide data from which the true height can be estimated. It may be of the conventional aneroid type, but the standard issue in the Royal Air Force has a scale which is not sufficiently open. In the course of a photographic flight the pilot can only give an occasional rapid glance at his aneroid and, unless the graduations are sufficiently widely spaced, he will not be able to notice small changes of height that may be important in the survey. In the latter part of our work we used a special aneroid which, though similar in many respects to the standard issue in the Royal Air Force, has a scale that is twice as open as the standard; it is also more delicately adjusted so that its indications do not depend appreciably upon whether it is rising or falling. When using the standard aneroid, we experienced considerable difficulty in getting the various strips of the mosaic at the same height; the aeroplane lost or gained height slightly during the turns between strips, and the height of the strips was influenced by the direction of approach to the desired height, whether from above or below. After we had obtained the special aneroid these difficulties practically disappeared.

In other chapters the opinion was advanced that an instrument should be designed which would adjust the throttle automatically in accordance with the air pressure, and thus maintain the aeroplane between definite limits of pressure, which would approximately correspond to definite limits of height. It will now be realised that such an apparatus might be of great value, even were it incapable of maintaining a constant height more accurately than the unaided pilot. The whole trend of the experiments which have been described indicates that the removal of any detailed operation from the responsibility of the pilot increases his accuracy in the remaining operations. It should, therefore, be clear, from the experimental data given in chapter IV, that such an instrument would be useful, provided that it could maintain height within 100 ft. of the mean value desired. There would appear to be no physical reason why a well-designed instrument should not give greater accuracy than this.

Let us now consider the problem of estimating the true height. The aneroid altimeter, being merely a pressure measuring instrument, is incapable of giving true height, unless the temperatures at all heights intermediate between the aeroplane and the ground are known. We saw in chapter III, however, that an estimate of true height, accurate within 1 per cent., can be obtained from a knowledge of temperature and pressure at the height of the aeroplane. The graduations of the service altimeter are calculated on the assumption that the air at all heights has a uniform temperature of 10 degrees centigrade, and that the pressure at the level above which height is to be measured is equivalent to the pressure of 760 millimetres of mercury. A very serious error therefore arises from the fact that the temperature of the air, in reality, falls rapidly with increase of height, thus causing the air to be more dense at high altitudes than it would be were the temperature 10 degrees at all heights. The difference of pressure between any two heights is therefore, in general, greater than is contemplated in the calibration of the standard British aneroid, and hence these instruments generally read too high at great heights.

It is a curious fact, discovered experimentally by the Research Department of the Air Ministry, that a single table of correcting factors, depending solely upon the air temperature on a level with the aeroplane will apply with very fair accuracy to all heights between 6,000 and 30,000 ft. These correcting factors are given herewith in Table 6.

Table 6. *Correcting factors for Standard British Aneroid Altimeters. Accurate within 1 per cent. of height for heights between 6,000 and 30,000 ft.*

Temperature on level with aeroplane. Degrees Fahrenheit	Height to be subtracted from aneroid reading. Feet
− 30	2100
− 20	1570
− 10	1145
0	804
+ 10	570
+ 20	302
+ 30	130
+ 40	0
+ 50	− 65

If a curve constructed from this table be used to correct the readings of an accurate standard aneroid, the error of the corrected reading will, in the British Isles at least, scarcely ever exceed 1 per cent. This rather surprising result was reached by analysing the temperature and pressure distribution for a great number of days on which pilot balloon ascents had been made, then calculating the errors in the readings of height which would have been given by a standard altimeter and plotting the results against temperature. Some modification of this curve might be required for use in tropical countries, but the changes would, at the worst, be slight.

The altimeter is subject to a further source of error, because it is graduated on the assumption that the air pressure at sea-level is 760 millimetres of mercury. This would be a serious error if allowance were not made for it by moving the scale, as a whole, relative to the pointer, until the reading on leaving the ground is either zero or corresponds to the known height of the aerodrome above sea-level. This procedure cannot give absolutely accurate results, but the error introduced by it is quite negligible, from the present point of view.

Still another source of error lies in the changes of pressure set up in the cockpit by the passage of the aeroplane through the air. This error might exceed 100 ft. in either direction, although it is unlikely to be so large as this unless the cockpit is very much enclosed; it could be removed entirely by making the case of the aneroid air-tight and connecting it to the static pressure tube of the airspeed indicator.

The aneroid altimeter itself is subject to instrumental errors, but these can be reduced to a negligible quantity by careful construction. The altimeter used in accurate work would naturally be carefully chosen and calibrated over the range of pressures and temperatures for which it will be used.

THE THERMOMETER. The thermometer we have used on the aeroplane is a standard issue of the Royal Air Force, designed to be carried on one of the wing struts, so as to be free from temperature disturbances set up by the engine. It must, of course, be graduated on a scale large enough to be read from the observer's seat. It is illustrated facing p. 112.

THE AIRSPEED INDICATOR. The measurement of air speed throughout our experiments has been made by the standard method in use in the Royal Air Force. Two tubes are exposed to the air, one of which is open at the end facing the wind and the other closed, but pierced by small holes in its cylindrical surface.. When two such tubes lie with their axes parallel to the relative wind, the pressure difference between them is closely equal to $\frac{1}{2}\rho V^2$, where ρ is the density of the air and V the velocity of the tubes relative to the air. These two tubes are called respectively the "pitôt" and "static pressure" tubes. The difference of pressure is indicated by means of a delicate manometer, which is graduated to read miles per hour in air of "standard density"—a conventional density approximating to the average density near the ground. The indications of this instrument will in general require correction, both to allow for any difference between the air speed past the tube and the speed of the aeroplane relative to undisturbed air, and to allow for changes of density consequent upon changes of height and climatic conditions.

In surveying, one important function of the airspeed indicator is to enable the pilot to maintain a constant speed, no matter what that speed may be. When it is used for this restricted purpose, the errors above mentioned are of no interest, because the work will nearly always be done at a constant height, and it is unlikely that the temperature changes during any one flight will be sufficient to introduce serious changes in the density. We have used the standard equipment of the Royal Air Force for this purpose, and found it to be adequate, but we think that a somewhat more open scale in the indicator might be adopted with advantage.

As far as possible we have avoided methods of surveying which require the accurate measurement of absolute air speed, for although easy enough in theory, this involves in practice a considerable amount of extra work of calibration, and the results are absolutely dependent on the avoidance of leaks in the tubes conveying the pressures to the indicator. When it is necessary to find the true air speed, it will be advisable to calibrate the instrument by actual trial. This can be done by flying the aeroplane at the required speed, on a fairly calm day, between two points at a known distance apart, flights being made in each direction several times in succession; the mean speed of all the flights in both directions is then taken to be the mean speed through the air, and is compared with the average readings of the instrument. This simple method is extensively used in experimental establishments to calibrate the airspeed indicators on aeroplanes. Its accuracy depends upon the wind remaining constant in strength between successive runs in opposite directions; it also depends upon the absence of a strong cross wind component, although some allowance can be made for this if the flights are along a straight road or railway, by reference to which the angle of drift can be estimated. The correction for cross wind, being proportional to the cosine of the angle of drift, does not become appreciable unless the drift is sufficient to be very apparent. In carrying out calibration experiments by this method, the aeroplane should not be too high, otherwise it may be difficult to fix the moment at which it is vertically over the known points.

An alternative and very successful method of calibrating the airspeed indicator is to lower a "static pressure tube" to a sufficient distance below the aeroplane to ensure that it is free from all disturbing influences. This method requires special apparatus, such as that developed for the purpose at the Research Department of the Air Ministry. Given correct apparatus, this is probably the best known method of calibrating the airspeed indicator. It is capable of giving very high accuracy.

Either of these methods will serve to calibrate the airspeed indicator for work in air of one particular density, but a further correcting factor will have to be used whenever the density of the air differs from that for which the instrument is calibrated. The principal factor that determines the density of the air is the height above sea-level and, for approximate work, it may be sufficient to assume that the density is the average density throughout the year at the height of the aeroplane. When this is sufficient an airspeed indicator which is calibrated for air of "standard density" can be corrected for height variations by multiplying the readings by the factors in the following table.

Table 7.

Height in feet	Correcting factor
0	0·988
5,000	1·069
10,000	1·159
15,000	1·259
20,000	1·370

The figures in this table, being averages for the year, will be liable to errors up to 3 or 4 per cent. on account of the daily and seasonal fluctuations of temperature and pressure. If greater accuracy is required it will be necessary to make a separate calculation for the correcting factor upon each occasion. This can be done by the method described in Appendix VI.

WIND GAUGE BEARING PLATE. This instrument, illustrated facing p. 157, can be used to find the strength and direction of the wind at the survey height and thence to calculate the bearing upon which the aeroplane should fly and the time intervals that should elapse between photographic exposures. It is a standard issue in the Royal Air Force, where it is employed in connection with aerial navigation. We have used it throughout the experiments described in chapters V and VI and we consider that it should be included in the equipment of any aerial survey. It is true that, in some circumstances, it may be possible to dispense with this instrument and to obtain the necessary information by other and simpler methods—as, for instance, by flying up and down the first strip without taking photographs—but there will be many occasions on which these simpler methods cannot be conveniently adopted. The advantage of the bearing plate method is that it can be used in any country in which it is possible to note a point passing under the aeroplane and keep it in view for two or three minutes; no pre-arranged land marks are necessary and the operation can be carried out wherever the observer chooses. Although therefore it is probable that simpler methods may occasionally be employed in special circumstances, we consider that the aeroplane should always be equipped with the wind gauge bearing plate or some equivalent apparatus, and that the crew should be trained in its use.

The wind gauge bearing plate which we have employed is illustrated and described in Appendix VII, where the methods by which we have used it are discussed in detail. The theory of this instrument is well known and is fully discussed in many manuals of aerial navigation, but we have judged it advisable to include this appendix, both for the sake of completeness and to show the particular method which we have found most suited to the problems in hand. For the present purpose it will be sufficient to remark that the methods we employ require the aeroplane to be flown straight and steadily by compass, on three successive bearings differing by 120 degrees, and that the whole operation of finding the wind and calculating the bearings and exposing intervals for the day's work can be carried out by a practiced crew in about 10 minutes.

We shall conclude this chapter with a brief description of the routine which we adopt for finding the wind, and for carrying out the other preliminary operations that precede a mosaic survey.

After deciding upon the ground to be surveyed, the pilot must be provided with some means of identifying the starting and ending points of each photographic strip. The true height at which it is desired to take the photographs must be settled, and the magnetic bearing of the strips ascertained. The observer should be provided with the following information:

1. A curve showing the compass deviations on every compass bearing.
2. A curve showing the aneroid readings which will be required at any temperature for a given true height.
3. A factor from which the time interval between the exposures can be derived from the length l* on the wind gauge bearing plate, or alternatively, if great accuracy is required, a family of curves from which this factor can be derived in terms of the altimeter and thermometer readings.

All these curves should be firmly fixed in convenient places about the

* See Appendix VII.

cockpit; on no account should they be carried loose in the aeroplane. They can generally be left permanently in an aeroplane during a survey.

The observer's cockpit should contain properly constructed pockets for carrying a slide rule and one, or better two, stop watches. One stop watch at least should be carried, even when the exposures are made by an automatic timing device, for this may get out of order and the observer have to carry on by hand. Finally the observer should carry two blank forms mounted on three-ply wood, or stiff paste board; one of these is

Observer		Flight No.....	Date ....
True Height	Temp.	Aneroid Height.	
Course Number		1	2
Mag. Bearing "B"			
Course to be flown "A"			
Corrected Compass Bearing			
Azimuth Reading		$A_1 - A_2$	$A_2 - A_1$
$T = \boxed{\dfrac{sd}{kv_1}} \times \boxed{\dfrac{b}{760}} \times \boxed{F} \times \dfrac{1}{\ell}$		Observer's Data	
$\boxed{} \times \boxed{} \times \boxed{F}$	ℓ		
$\boxed{} \times \boxed{}$	T		
$\boxed{}$	Drift Angle	$B_1 - A_1$	$B_2 - A_2$

Fig. 2. Specimen form for Observer

for his own use, and contains spaces for him to enter the results of his calculations and clear instructions which leave nothing to be thought out in the air; the other is to be filled in after the calculations are finished, and handed to the pilot for his information. Both pilot and observer should provide clips in which these forms can be held in positions from which they can easily be read. Convenient arrangements of these forms are illustrated in figs. 2 and 3. These are not precisely similar to the forms which we used, but the general principle is the same. On the full size, the longer dimensions of the forms illustrated would be about 6 inches. It is probable that every surveyor will favour his own type of form;

those illustrated are merely specimens of the type of form which appeals to us. The important point is that some adequate form should be carried by pilot and observer.

After reading the descriptions which have been given in the text and in the appendices, of the operations that have to be carried out in the air, these forms should explain themselves. It will be sufficient to state that they are drawn out to be suitable for use with mosaic or navigation methods, in which two courses, differing by 180 degrees, have to be flown. "B" is the magnetic bearing of the course to be made good over the ground. "A" is the true magnetic bearing upon which the aeroplane will have to fly in order to make good this course. The "Corrected Compass Bearing" is corrected for compass deviation. The "Azimuth Reading" is the reading which should be given by the gyro azimuth

Pilot.................	Flight No......	Date........
Aneroid Height.		
Courses.	1	2
Corrected Compass Bearing.		
Azimuth Reading.		

Fig. 3. Specimen form for Pilot

indicator, after it has been set to zero near the end of one of the courses and used to turn onto the next course. "T" is the time interval between exposures upon each course. The "Drift Angle" is the angle through which the camera will have to be rotated in the aeroplane in order to prevent the photograph from being twisted relative to the track of the aeroplane. Space has been provided for calculating "T" from the longer and more accurate formula given in Appendix VII. The method of carrying out this calculation, when in the air, is discussed in that appendix, but it is to be noted that it will only be necessary to use this rather elaborate process when great accuracy is required. The observer would leave the ground with the spaces for "True Height," "B," and the first three rectangles in the calculation for "T," already filled up. The figures that he will require during the photographic work are surrounded by specially thick lines. The pilot's form explains itself.

After ascertaining that all the apparatus is in place, and after all the calculations that can be made before leaving the ground have been completed, the pilot begins by climbing to the approximate height of the survey, whilst flying on a course which will bring the aeroplane to a point some distance to windward of the starting-point of the first photographic flight. On reaching this height the climb is stopped and the aeroplane flown level for a short time, to allow the thermometer to settle down to the correct reading. The observer then reads the temperature, enters it on his form, looks up the aneroid reading that corresponds, at this temperature, to the required true height, enters this reading on his form and communicates it to the pilot. The latter then ascends or descends, until the aneroid registers this given reading, and begins to trim the tail and adjust the throttle until the aeroplane will fly by itself, upon an even keel, at the required height and speed. During this time the observer can be calculating, from the readings of the altimeter and thermometer, the value of the factor that he will require later in the determination of the time interval between exposures; if he has any more spare time, he gives the camera a preliminary adjustment for tilt.

When the pilot is satisfied with the adjustment of his aeroplane, he signals that he is ready to start on the wind finding triangle, and sets his aeroplane onto the first of the pre-arranged courses; the observer sets the glass disc of the wind gauge bearing plate, as explained in Appendix VII, and waits until the aeroplane has been on the given course for about two minutes, before he starts to measure drift. He measures drift by taking "tail bearings"* on points which have passed immediately beneath him, using, as a rule, two or three points in succession for each bearing. If the pilot is flying well, the observer should not require more than two or three minutes to satisfy himself as to the correct setting of the drift bar, so that the first course will occupy from four to five minutes altogether.

Towards the end of this straight flight, and before diverging from his course, the pilot releases the gyro azimuth indicator in readiness for the turn. On receiving a signal from the observer, he turns through 120 degrees by the azimuth indicator and starts upon the second side of the wind finding triangle. Since the gyro azimuth indicator, unlike the compass, is quite unaffected by the turn, no time need be lost on this course in waiting for the compass to settle down; the observer can therefore start straight away on the measurement of drift, which should not take more than two or three minutes. A third turn and a third measurement of drift complete the process.

If the gyro is working well, it should not be necessary to readjust it on the second course, preparatory to the second turn, but the pilot should know the corrected compass bearings for the second and third courses, in order to verify that the gyro is not precessing badly.

* See Appendix VII.

The whole of this operation of finding wind, including the subsequent calculations, can be performed by a practiced crew in 10 minutes when the gyro azimuth indicator is used, but if this instrument is not available at least four more minutes will be required to allow the compass to settle down on the second and third courses. The gyro rudder control is not essential for this operation, but when it is available it can be used with advantage to hold the aeroplane straight on each course.

When the wind triangle has been completed, the pilot sets a course which will carry him to a suitable spot from which to approach the first starting-point of the survey. The observer meanwhile carries out the necessary calculations, filling up his own and the pilot's forms and handing the latter to the pilot. He then sets the exposing mechanism to the calculated interval for the first photographic flight and gives the camera the necessary rotation in azimuth to allow for drift. He gives the camera a final adjustment for tilt, by reference to the spirit levels, and awaits the moment for the first exposure. The pilot meanwhile, having reached a suitable point from which to begin operations, lays his aeroplane upon the calculated compass bearing; if he has manœuvred skilfully, about two minutes on this course will bring him close up to the starting-point of the first photographic flight, in such a way that only a small alteration of course is needed to cause him to pass over it. The compass should by now have settled down so that he will be flying truly on the required bearing, and he may therefore throw the gyro azimuth indicator into action, ready for use on the photographic flight. Having done this, he concentrates upon getting accurately over the first starting-point and if, whilst doing so, he gets off the correct course, he immediately turns until the gyro azimuth indicator again reads zero. If he has a gyro rudder control he now throws it into action and proceeds on the photographic flight.

During the photographic flight itself neither the pilot nor the observer should bother at all about the ground beneath them, but should concentrate upon the work in hand within the aeroplane. This instruction, important in any case, is doubly important when the gyro rudder control is not used, for then the quality of the flying is liable to deteriorate, if the pilot so much as moves from his position of sitting squarely in his seat with his eyes level. Even the observer, if he moves about much, or exposes his head and shoulders to the air blast, will seriously increase the pilot's difficulties. When not using the gyro rudder control, the pilot should keep straight by reference to distant objects near the horizon, or to sun shadows in his cockpit; he should not try to fly entirely by compass.

On approaching the end of the photographic flight the pilot verifies that the azimuth indicator is reading zero and, if necessary, readjusts it by locking it in the zero position and releasing it again; this is to prepare it for use on the turn to the return flight. On receiving the observer's signal that the photographic flight is ended, the pilot looks down and identifies the pre-arranged ending-point over which he should be and, if his course has not carried him directly over it, he makes a note to correct the subsequent courses accordingly. A rough estimation of this correction is easily made from the consideration that on a 10 mile flight an error of one degree corresponds to about one-sixth of a mile. The observer at the same time will note any error in the timing of his exposures, and record a corresponding simple correction. The pilot now turns through the required number of degrees on the azimuth indicator and gets over the starting-point of the first return flight, whilst the observer alters the time interval between exposures and the allowance for drift on the camera. The routine is then continued until the day's work has been completed.

It will have been noticed that the compass is never used except to adjust the gyro azimuth indicator in preparation for the first flight and, towards the end of each photographic flight, to readjust it in preparation for the next flight. If no gyro is used, it will be necessary, in order to allow the compass to settle down, to hold the aeroplane upon the required bearing for some two minutes before the photography is started in each flight; and to obtain room to do this the pilot will have to fly on before turning for some two minutes after each flight. At least four minutes will thus be wasted between each photographic flight and more than 50 per cent. will be added to the time required to survey a given area. If the amount of work that a pilot can do in a day is limited by the time that he can spend upon the actual surveying, it follows that the use of the gyro azimuth indicator will make it possible to cover some 50 per cent. more ground for a given amount of flying. The gyro rudder control on the other hand does not affect the amount of work that can be done in a given time, but it relieves the pilot of a very exhausting operation and thus allows him to work longer hours in each flight. Both these gyroscopic instruments should therefore be of great economic value in aerial surveying, by increasing the area of ground which can be covered in a single flight; this, of course, is quite apart from the improvement that they will certainly make in the quality of the work.

The operations involved in surveying by navigational control are similar to those we have described for mosaic work but they are somewhat simpler. The wind at the survey height has to be found as before, and the flight has to be started in a given direction over a given point, but the flights are longer and more widely spaced, so that great accuracy in getting over the starting-point is not so essential and the operation is repeated less frequently.

CHAPTER VIII

TRAINING

THROUGHOUT the previous chapters great emphasis has been laid upon the necessity for training both pilots and observers for the special requirements of aerial surveying. The general character of the training required has, perhaps, been indicated when discussing the various problems which have to be faced, but it is proposed in this chapter to specify, as far as possible, the precise nature of the training which we consider should be given to pilots and observers who are going to carry out these methods.

Before coming to the detailed training, it is necessary to emphasise the importance of employing none but good and experienced pilots; this is not so much on account of the technical difficulties of the survey itself, as on account of the extreme importance of avoiding accidents. The flying may often have to be carried on under difficult conditions, with the possibility of forced landings on unsuitable ground, and no commercial undertaking could long survive the losses resulting from an undue proportion of accidents to personnel and equipment. Under no circumstances therefore should the organiser yield to temptation, and employ pilots who have not had a wide and varied experience of flying.

For the observer an extensive flying experience is not necessary, but, even in his case, it is useless to expect results from a man who has had no experience of flying. It is essential that the rather complicated operations which fall to the observer should be carried out without mistakes, and the difficulty of performing in flight precise operations without mistakes must be experienced to be understood. A large number of influences—cold, fright, lack of oxygen, noise and lack of room, combined with heavy clothing—conspire to distract the observer's attention and to reduce his powers of precise detailed thinking. A long experience of experimental aeronautics has shown that very few observers are capable of doing accurate work upon their first few flights. Indeed, some people appear to have temperaments that unsuit them permanently for aerial observing, but it is found that men with suitable temperaments only require a few practice flights before they are ready to engage in accurate work.

It is assumed, then, that the pilots to be trained are already experienced and skilful in the ordinary operations of flying, and that the observers have had a few preliminary flights of moderate duration, at approximately the height at which the work will be done. From the very beginning of the specialised training, pilot and observer should work together, the pilot attempting to carry out certain flying operations as accurately as he can, and the observer providing a positive check on his work, by means of photographs. It will be found that the best results will be attained after individual pairs have been working together for some time, but it may not be possible to carry out the work in the field with the same pairs always working together, hence, in the training, the combinations should be changed from time to time. In the field, however, it may be worth while to go to considerable trouble to keep the same pairs working together.

Speaking generally, the equipment for training should be the same as that which it is proposed to use in the field, but this is not absolutely essential, and it may be advisable to vary the equipment somewhat, with a view to carrying out the training more economically. If it is impossible to train in the aeroplane which will be used in the field, it should, at least, be insured that the training aeroplane can climb to 10,000 ft. without difficulty, and that it should have approximately the same characteristics as the aeroplane to be used in the field. For example, training for work upon a tractor should not be carried out upon a pusher, and the training aeroplane should, at least, be stable at its cruising speed and have controls which can be trimmed in flight.

The pilot's compass should be of the same type as the one he will use in the field; this is most important. As far as possible he should have the same instruments arranged in the same way as in the aeroplane in which he will ultimately be working, but it is advisable, in the early stages of his training, not to let him use either the gyro fixed azimuth or the gyro rudder control. The observer should have a "vertical" camera which might either be the one he is going to use in the field, or, if this is too expensive in film etc., a cheaper, non-automatic camera, such as the L.B., would serve the purpose. Whatever camera is used, it must include some arrangement for recording, on each photograph, the optical centre and at least one datum line.

The observer will require a wind gauge bearing plate, or its equivalent, and he must either be provided with a good open scale aneroid, or be able to see the pilot's aneroid without effort. Some means of communication, such as speaking tubes, should be provided between the pilot and observer, and they should be close enough together to allow of written notes being passed from hand to hand.

It is of great importance to choose a suitable district for training. An ideal district would be moderately flat, contain much previously mapped detail that can be identified on photographs, should be good for forced landings and, above all, enjoy suitable weather. The district round

Cambridge is ideal in every respect but that of weather, and in this respect it is hopeless; it is, in fact, probable that there is no part of the British Isles in which the weather is suitable, either for training, or for aerial surveying on a commercial basis. The work cannot proceed unless the sky is practically free from clouds, and it is most important, both from the point of view of economy and efficient training, that the work should be moderately continuous.

It is an important feature in the training that the pilot and observer should be able to study one day's work before proceeding to the next, and it is therefore necessary to provide for the rapid development and printing of the photographs, and for the identification of their positions upon the map. We have found that good one-inch maps, such as the British Coloured Ordnance, or the Ordnance "Popular" maps, are suitable for this purpose. Smaller scale maps are of little use, whilst larger scale maps, such as the six-inch, are awkward, because of the number of separate sheets covered by a single flight.

Before starting on the practical part of the training, it is advisable to make sure that both pilot and observer understand the main theoretic principles which govern the methods they are to employ. They must, for instance, understand the danger of relying upon the apparent vertical. They must know the main causes of error in the compass, that it must not be used at all until the aeroplane has been flying straight and steadily for some time, that it is affected by turns on North-South courses and by changes of speed on East-West courses, and finally that it is more difficult to use it when flying towards the magnetic Pole than when flying in any other direction. Within limits, the more that the pilot and observer know about the theoretic principles of their subject, the better will they be able to profit by experience in their training. They should at least be familiar with the principles enunciated and discussed in the preceding chapters.

Assuming that the observer has had previous flying experience, the practical part of the training might begin by the pilot and observer ascending to a pre-arranged aneroid height, and there carrying out four practice flights on different compass bearings, say North, South, East and West by compass, without correction for deviation or declination. Each of these flights should be about 10 miles in length and should be as straight and steady as possible, the speed being kept constant at some convenient pre-arranged pitôt reading and the height maintained, as accurately as possible, at a pre-arranged aneroid reading; for example, 10,000 ft. It is essential that this practice should be carried out at a sufficient height to be clear of turbulent air.

During these flights the observer should take photographs at accurately timed equal intervals of about half a minute, beginning not less than one minute after the pilot has taken up his course. If he is using the L.B. camera, he will use up the eighteen plates in one box in $8\frac{1}{2}$ minutes, so that each flight might conveniently continue for this length of time. The observer should record the reading of his aneroid between each exposure, unless this record appears automatically upon the photograph. The camera during this practice should not be adjusted, but should remain fixed relative to the aeroplane throughout the flight.

The object of this exercise is to train the pilot in straight steady flying at a constant height, and to check the results by means of photographs; at the same time accustoming the observer to the routine operation of exposing a camera at regular intervals. It is probable that, on actual survey, the operations of film changing and exposing at equal intervals will be automatic, but it is of great importance that the observer should be able to throw out the automatic gear and continue by hand if necessary, and it is therefore advisable to accustom him, from the very start of his training, to work by hand. The automatic gear, when available, should be looked upon as a luxury, well worth having when it can be obtained, but not absolutely essential to the success of any particular flight.

To study the results of this exercise we observe that the accuracy with which the pilot has been able to maintain a constant height can be ascertained from the observer's records, and that the quality of the flying can be studied by plotting the optical centres of the photographs upon a map and noting the degree to which they are equally spaced along a straight line. To obtain a quantitative measure of the accuracy of the flying, the "local" errors should be determined, as in chapter VII, by finding the displacement of each optical centre from the point that lies midway between the centres of the photographs immediately before and behind it. This affords a very complete test of the quality of the flying, because good results can only be obtained by flying in a straight line, at a constant speed, without varying either lateral or fore-and-aft tilt. Finally, the general straightness and steadiness of the flight can be studied by joining the centres of the first and last photographs of each flight and noting the displacement of the centre of the middle photograph of the flight from the midpoint of this line*.

The errors that are permissible in this practice can be judged from

---

* If this error is large, but is thought to have been due to a change of wind during the flight, the question can be settled by comparing the photographs with the map to find their azimuth orientation. This process is, however, relatively laborious and requires good maps; it should seldom be necessary to apply it. If the number of exposures has been even, so that there is no middle photograph, discard the first or last photograph.

the results of our experiments recorded in chapters IV and VII. They may be stated somewhat as follows. The difference between the greatest and least heights recorded in all the four flights should not exceed 200 ft.; the 50 per cent. deviation from the mean height of all four flights should not exceed 50 ft. The "local" errors should seldom exceed 0·1 mile; the 50 per cent. "local" error should not exceed 0·04 mile. The error in the test for general straightness and steadiness of the flights should seldom exceed 1·5 per cent. of the length of the flight and should, as a rule, be under 1 per cent. If the results are appreciably worse than is indicated by these figures, more practice will be required before proceeding to the next stage in the training. It is thought that a pilot of average skill should not require more than two or three such practices before attaining this accuracy.

When the crew have passed this practice, but not before, they should proceed to a second stage, which should include the operations of finding the wind, and of flying over a pre-arranged point whilst on a given compass course. For this practice the observer will require, in addition to his previous equipment, a wind gauge bearing plate; whilst the pilot might, with advantage, be allowed to use the gyro fixed azimuth to supplement his compass.

It is probable that the best way of obtaining practice at these operations is to carry out a small skeleton survey on the lines suggested in chapter V. Two parallel lines, about 10 miles apart, might be ruled on the map and be marked off by a series of points spaced about a mile apart; these would be the starting and ending points of the separate flights in the skeleton survey. The crew would be instructed to find the wind at a pre-arranged aneroid height, and to calculate the compass bearings which will carry them in the required directions, when flying there and back between these points. They should then carry out six or eight flights of a mosaic survey, as described in chapters V and VII, but with the following exceptions: to save plates and to avoid giving the observer too much work at this stage, he might take photographs at intervals of two or three minutes, instead of the shorter intervals, which would be necessary to obtain an overlap; these should be sufficient to indicate the straightness and direction of each flight, and the observer would be free from the necessity for working out the exact time interval between exposures. No attempt should at this stage be made to adjust the camera either for azimuth or tilt. By omitting all these operations the observer is relieved of much of the heavy work of a real mosaic survey, which he will probably not be competent to perform at this stage of his training. He should, however, be instructed to record aneroid height at intervals, in order to check the pilot, a thing which he will not be re-

quired to do in a real mosaic survey. The pilot should be instructed to
fly always on the calculated courses, making no attempt to alter course
between successive flights, as he would in a real survey should he find that
his flights are not ending at the correct points.

Much information can be obtained from a study of the results of this
practice. In the first place the distance between the path of the aeroplane
over the ground and the pre-arranged starting-points over which it
should have passed can be measured; with good flying this distance
should only occasionally subtend more than four degrees at the height
of the aeroplane. Next, the extent to which successive flights in one
direction are out of parallel can be estimated The angle between any
two strips that are intended to be in the same direction should not
exceed some three degrees, and the strips should be straight, within the
limits laid down in the previous practice. Again, the mean bearing of all
the strips in one direction should not differ from the proposed bearing
by more than three degrees. Finally, the height should have been kept
within 100 ft. of the mean height of the whole series of flights.

If the records on the wind gauge bearing plate have been preserved—as should always be done—it will be possible to obtain some
idea of the accuracy with which the wind has been found, from a
consideration of the size of the triangle formed by the three drift lines
on the plate—see Appendix VII; but this is not a very reliable guide,
particularly with inexperienced observers, who are apt to allow their
judgment upon the position of the third drift line to be biased by
their desire to obtain a small triangle of error. It will always be advisable
to repeat, after he has returned, all calculations made by the observer
in the air, to ascertain whether such errors as exist occurred in the
calculation, or are due to errors in the observations.

When this second practice can be carried out in a satisfactory manner,
the final practice of constructing a complete mosaic of about 10 miles
square should be attempted. This should be carried out exactly as
described in chapters V and VII. The operations will be much the same
as in the previous practice, but they will extend over a longer time and,
on the part of the observer, will be much more exacting. The work
should be carried out at a pre-arranged absolute height, which will
involve correcting the aneroid in accordance with temperature; the full
number of photographs should be taken with the correct overlaps and
with the camera continually adjusted for drift and verticality. The
results should be worked up into a finished mosaic, the excellence of
which might serve as a kind of certificate of competence for the crew.

Up to this stage the gyro rudder control has not been used in the
training, and, in its absence, the crew will not be able to obtain results

of the greatest possible excellence, but it is considered to be important that they should be trained to make a passable mosaic without it. It will be found that pilots who can produce a good survey without the rudder control, will produce better work, with less fatigue, when using the control. A crew who can produce a passable survey of one hundred square miles without the gyro control may be considered to be fully trained; none of the other problems which they will have to face will be so difficult as this.

It will be noticed that we have not suggested that the pilot be required to carry out a mosaic survey without the help of the gyro fixed azimuth; this is because we have not carried out a survey in this way ourselves, and we are not quite certain how difficult it would be, particularly when working from a tractor. As a matter of fact we have had practically no trouble with this instrument, during the whole of our experiments, whereas we have had considerable trouble with the rudder control. We are inclined to think that it would be possible to do without the azimuth indicator when working from a pusher, but that, when working from a tractor, its absence would be severely felt. In any case we consider that this instrument should be used whenever it can be obtained.

It must not be supposed that the pilots employed in our experiments were trained in the systematic way above suggested. Their experience grew up with ours during the experiments.

Major Griffiths' fatal accident occurred before we had an opportunity to discuss in detail the question of routine training, so that this scheme has been drawn up without his approval, or the benefit of his assistance. This is unfortunate because Griffiths had had the bulk of the flying experience. After his death I had, however, a considerable experience of observing in the experiments on Navigational Control, and before his death we had had many talks on the matter of training. I believe that he would have agreed, at least with the general outline of this scheme, even had he wished to modify it in detail.

CHAPTER IX

CONCLUSIONS

IT is proposed in this chapter to review the bearing of the experiments which have been described upon the question of applying aerial methods to practical surveying. It must first be stated that, in making these experiments and in writing this book, we have neither desired nor attempted to prove that any one method of aerial surveying is more generally useful than any other; nor indeed that the aerial method itself is, in general, a more satisfactory way of constructing maps than the older methods of surveying on the ground. In practice the proportions in which the aerial and ground methods can best be used, and the decision as to the precise aerial method to be employed, will depend entirely upon the circumstances of each particular case. The director of the survey should naturally be thoroughly conversant with the details of all the methods which are available to him, and he it is who will have to decide the extent to which any method should be used in each instance. He cannot, however, expect to reach a correct decision unless he is provided with data concerning the accuracy to be expected and the difficulties to be encountered with each method which comes under his review; it is with the object of providing some of the data, relating to the aerial side of the problem, that the experiments were made and the book written.

Even on the aerial side, however, the book does not claim to be a complete exposition of the problem, as it is at present understood, for it does not, except in one particular instance and for a specific purpose, deal with the group of methods which may be classed under the general head of "re-section," and little reference has been made to any of the very useful methods that have been developed elsewhere, for the analysis and interpretation of oblique photographs. These omissions must not be taken to imply that we hold the opinion that other methods are less valuable than those which we have studied. The experiments were originally undertaken because there appeared to be a definite lack of precise information in certain directions, and the book has been written to record the results of those experiments and to discuss the developments which the experimental results indicate should be realisable. We are of the opinion that there are many surveying problems to the solution of which one of the variants of the re-section method will be more suitable than any of the methods we have studied; we also hold the view that there are many other circumstances under which the re-section methods will not be applicable, but for which the methods we have studied, or similar methods, will prove to be of the greatest value.

With this explanation we shall attempt a short summary of the argument running through the book and of the principal conclusions to be drawn from the results of the experiments.

With the single exception of surveys in which economy of flying time and photographic material is a matter of no importance, the mere problem of arranging that all the ground shall be adequately photographed without leaving gaps is one of considerable difficulty, requiring for its solution the employment of adequately trained pilots, suitably equipped and working on a carefully thought-out plan. The difficulties of this problem are of a nature which are not likely either to be understood, or even realised, by a pilot who has neither had his attention drawn to them nor made a special study of the matter. This preliminary difficulty alone puts aerial surveying into the class of operations which must be performed by specialists.

We feel from our experience that we cannot lay too much emphasis on the statement that to attempt a large aerial survey, without specially instructed or experienced pilots and observers, is to court almost certain failure from the economic point of view; at least in the early stages before the crews have discovered, by personal experience, the various pitfalls which have to be avoided. We think that this will be the case whatever the methods employed. On the other hand we are definitely of the opinion that the work does not require exceptional inherent skill on the part of either pilot or observer, but rather that it could be adequately performed by any crew of ordinarily capable people, provided with the necessary equipment, instruction and experience. We have attempted in chapters VII and VIII to indicate what this equipment and training should be. If our opinion is correct, aerial surveying becomes exactly comparable with most other routine engineering or scientific activities, for these nearly all require special equipment, instruction and training, but are not confined to men of exceptional ability.

In relation to this problem of covering the ground, which resolves itself in practice into the problem of flying in a series of parallel straight lines with a predetermined lateral spacing, our experiments and conclusions are of general interest. The remainder of our experiments, in which the basic idea is that the aeroplane should be flown with the height and orientation as constant as possible, apply more particularly to those rapid methods of surveying in which re-section from known ground points is not used; they have therefore only a secondary interest in connection with those methods in which it is used.

Two methods of rapid surveying based upon accuracy of flying have been studied. The first of these may be called the "mosaic" method, because the map is made by arranging the prints from "vertical" photographs, in such a way that the detail of the ground runs continuously from print to print, so that a kind of mosaic is formed. In circumstances where it is applicable this method should have great practical value; for, by it, maps containing an extraordinary amount of detail can be made at the expense of one linear mile of flying per square mile of country mapped, so that it should be possible to cover an average of about one hundred square miles of country in a day's work of reasonable duration. Similar processes were extensively used during the war, but, for the most part, were applied to districts for which a close ground-surveyed control existed. It is now shown that, except in cases where such a close control is available, the success of the method depends absolutely upon securing very accurate flying, such as can be obtained from trained pilots, but is not likely to be obtained without a certain amount of experience and special instruction.

It is recommended that mosaic mapping should be carried out in a series of ten-mile straight flights, about ten of which would cover an area of a hundred square miles and form an average day's work. It is further recommended that each day's work should be compiled into a separate mosaic and it has been shown, by actual demonstration, that these mosaics need not contain "local" errors—that is displacements of any one print relative to its neighbours—of more than some 0·02 mile, and that they can be made very free from systematic distortion, even when there is no ground control whatever. In the absence of any independent ground control, difficulty may be experienced in fitting the separate mosaics together and in locating each mosaic in its correct position in relation to some larger scheme; it was shown, however, that there is no serious difficulty in manipulating the mosaic to fit any independent control which may be available.

From these considerations it is deduced that the mosaic method is suitable for use with a ground control in which the surveyed points are separated as widely as 10 miles, but that a closer ground control, if it can be provided, would give increased accuracy. With a ten-mile spaced ground control, it was shown by actual demonstration that the 50 per cent. error* of position of any point can be kept down to about 0·03 mile and that the errors need never exceed some 0·07 mile; errors of the latter magnitude occurred on the extreme edge of the experimental mosaic, owing to causes which should not commonly occur when practiced crews are working to a routine. With a close ground control, say

* The 50 per cent. error is that error which is equally likely to exceed, or be exceeded by, any individual error. See Appendix I.

of one mile spacing, the errors would, of course, be less than these and should in fact be reduced merely to the "local" errors, having a 50 per cent. value somewhere in the neighbourhood of 0·01 mile.

It is to be noted that, when applied to country which has not previously been mapped, the mosaic method requires a certain small amount of preliminary flying, to provide information which the pilot can use to locate the starting-points of his mosaic strips.

This mosaic method, useful though it may be in certain circumstances, suffers from severe limitations. Most important of these is that it can only be applied to flat, or gently undulating, country. The accuracies given above can be reached in country in which the local changes of height do not exceed some 5 per cent. of the height of the aeroplane, or about 500 ft. when working from a height of 10,000 ft. In country which is more hilly than this the method might still be applied, but the inaccuracies would be greater and the labour of compilation of the map would be seriously increased. In country in which the differences in local height exceed some 10 per cent. of the aeroplane's height, we consider that the method could scarcely be expected to give more than a series of pictures of the ground, which might be of great value in indicating the general nature of the ground, but which could not be directly compiled into anything resembling a map. Again, the method, as at present conceived, will not, in itself, allow the absolute height of any point to be determined with useful accuracy; although it is known, from independent experience, not our own, that the steeper ground slopes can be sketched successfully, by viewing successive overlapping photographs in a common stereoscope.

The second method which we have studied, that of surveying by oblique photographs and what we have called "navigational control," has certain limitations of its own, but is not subject to the same limitations as the mosaic method. It is suitable to any kind of country, however hilly, and is capable, under favourable conditions, of giving some indication of absolute height. It can also be carried on more rapidly than the mosaic method, for it will give moderately detailed maps at the expense of one linear mile of flying to some five square miles of area mapped, that is to say at the rate of some 500 square miles of country mapped to an average day's work in the air. Finally, although a close ground control would always be advantageous, this second method can be applied when the ground-surveyed points are spaced as widely apart as 50 miles or even more.

This method would not result in a complete picture map, such as is given by the mosaic process, and when carried on at the rate above mentioned, would not give such detailed information concerning those parts of the ground which appear in the oblique photographs only; the latter defect could, however, be removed at the expense of more flying per unit area, which would still leave the rate of progress greater than in the mosaic method. In general we think that the navigation method should be looked upon as an alternative to the mosaic method rather than as competitive with it; nevertheless the fact that it can be carried out more rapidly and economically than the mosaic method may lead to the use of it, even in cases where the mosaic method is applicable.

The general idea would be to carry out a series of long straight parallel flights spaced rather widely apart—some 10 miles has been suggested—and to view the country between by means of oblique photographs which include the horizon. It is supposed that two such oblique and one vertical photograph would be taken simultaneously, and it has been shown that, using wide angle lenses, these three cameras could be arranged so that the whole of the ground would eventually be included in the photographs, at sufficiently short range to allow a reasonable amount of detail to be detected.

The map would be constructed by a method similar to that used with the "plane table" on the ground. The position of the aeroplane at the instant of exposure would be fixed primarily by navigation, and the photographs would be used to allow rays to be drawn from these positions towards the various ground features, which would eventually be located at the intersections of a number of such rays radiating from different points. The positions of the aeroplane at the moments of exposing the photographs would be fixed on the assumption that at least two known points occur on its path, and that it has proceeded over the ground between these two points in a straight line and at a constant speed. Various considerations have led to the suggestion that it will generally be advisable to carry out two series of parallel flights which would be mutually at right angles, and hence would form a kind of net over the ground, having a rectangular mesh of, say, 10 miles to the side; if this were done it would not be necessary to provide two ground-surveyed control points for each straight flight, because the flights would control each other to some extent.

A difficulty arises, in that it is impracticable to arrange for a straight flight, of the nature required, to pass closely over more than one *predetermined* point. Some of the control points would therefore have to be fixed after the photographic flights, or else the flights would have to be located by reference to control points lying at some distance to one side of their track. This difficulty has been discussed and the conclusion reached that it would not be impracticable to adjust the photographic net to fit a control consisting of even a primary triangulation, laid down without reference to the needs of the aerial survey. This, of course, is subject to the control points being visible from the air at a distance of a few miles. An independent triangulation would, however, be the most difficult form of control to use and, whenever possible, it would be preferable to lay down the ground control after the flights had been completed, so that the control points could be advantageously placed in relation to the positions of the flights which have occurred.

We have not been able to test out this method, as we did the mosaic method, by the construction of an experimental map, because our resources were quite inadequate to deal with work on the scale that would have been required. We have, however, made an experimental study of each of the processes involved, partly to detect unexpected difficulties, but mainly to provide data from which the accuracy of the resulting map may be estimated and the amount of detail which it will show ascertained.

Long straight flights, up to 90 miles in length, have been flown and studied; and oblique photographs, including the horizon, have been taken, with a wide angle lens, from a height of 10,000 ft. These experiments have led to the conclusion that, with a ten-mile spacing between successive flights, maps could be made in which most of the detail of the ground could be identified, with the exception of small clearings surrounded by high trees and situated between the flights. A closer spacing between the flights would allow more detail to be recorded but would of course require more flying to the square mile.

The 50 per cent. "local" error of these maps should be about two hundredths of a mile and the 50 per cent. "total" error should be between one-half and one per cent. of the distance between control points, provided that this distance is not greater than about 50 miles. It should be possible to determine the height of any well-defined point above sea-level with a 50 per cent. error of about 100 ft., and steep ground slopes could be sketched in by viewing successive overlapping photographs in a simple stereoscope. Occasional errors greater than twice the 50 per cent. errors must be expected, but the presence of these, when they occur, should be indicated to the surveyor by the inconsistency of his data.

The figures relating to height and "local" error assume that the visible horizon will appear in at least one of the three simultaneously taken photographs, and that the horizontal plane can be estimated from the visible horizon with the accuracy determined in England, by a series of Air Ministry experiments, to which reference has been made. When the horizon is not visible it becomes impracticable to determine absolute height with useful accuracy, and the "local" errors must be expected to increase to about double the above magnitude.

Let us conclude by a short discussion of the way in which we think that these methods might be applied to the successive surveying of a country, as it progresses from the unexplored to the civilised state. In the early stages it is to be presumed that no detailed surveys of any kind will exist and that the country will either be entirely unsurveyed, or a very open triangulation only will be in existence. In such a country some very complete system will have to be devised to enable the pilot to find his way and locate himself with sufficient accuracy to make sure that he is photographing the desired parts of the country. It is in these circumstances that the long straight navigationally controlled flights which we have described will be of great value, and we therefore consider that the natural procedure would be to carry out a preliminary reconnaissance survey by this method, whatever process is to be used in the construction of the final map.

The situation and sequence of the navigationally controlled flights would, of course, depend to some extent upon the nature of the country; they might, for instance, be made to span the region between navigable rivers, or go from shore to shore, in the case of an island; but it will be of more general interest to discuss the case where there are no special features to aid in the lay-out of the scheme, so that we must suppose that there are no well-known features to guide the pilot, and that he has to rely upon methods which will automatically ensure that he covers the whole ground. Under these circumstances it is probable that the best way of starting the survey would be to carry out four straight photographic flights, forming the four sides of a square of from 30 to 50 miles to the side; the form of the square being determined by purely navigational methods. The pilot would naturally turn outwards, through 270 degrees, at each corner, so as to cross his previous flight and thus ensure that the square is constructed entirely from intersecting straight flights. If three cameras are used in this flight, as has been suggested, the photographs would form a very good preliminary guide to the whole of the country in the interior of the square, and the strips of vertical photographs could be given to the pilot to take with him in the air to act as a guide in starting subsequent flights.

The next step would be to cross this square in a series of parallel flights starting from predetermined points on the edges; these points could be given the desired spacing on the assumption that the aeroplane during the previous flight had proceeded at a constant known speed along each side of the square. These spacings would of course be subject to errors of 3 or 4 per cent., corresponding to errors in the estimated ground speeds of the aeroplane in the first flight, but these variations would ultimately be checked by the ground control and would, therefore, be of no consequence, unless they rose to such high values as to upset the general outlay of the work.

Proceeding in this way, the whole interior of the square could be covered by a net of intersecting flights and the photographic material would be available for the construction of a map such as we have been discussing. It is to be noted that the whole of these operations could have been carried out without any help from pre-existing maps, or in fact without any co-operation with the ground. This is a point which we think will be found, in practice, to be of very great importance.

The construction of accurate maps from the photographs will, of course, require co-operation with some ground control, to which the net of flights must be adjusted. If the ground survey is already in existence, methods will have to be devised for straining the net to fit it; this, as we have stated in a previous chapter, could, we think, be carried out by representing the net elastically and straining it to fit the control, so that the proportional spacing between photographic plumb points, which is determined by timing during the flights, would be preserved.

In the case where the ground control has to be provided, the natural procedure would be to fix control points as near as possible to chosen intersections between the flights. This might be done in the usual way by triangulation, but should triangulation be difficult, we are informed that it is now possible to fix position by astronomical observations taken from the ground, with an accuracy of the order of about 0·1 mile. This method of controlling the aerial survey might prove to be very convenient, because it would make it possible to choose the control points with reference to their visibility from the air, and without reference to their visibility from other points on the ground.

It might well happen that the reconnaissance map constructed in this way would be sufficiently detailed and accurate for the early stages of the development of the region; particularly if care were taken to reduce the spacing between at least one set of the parallel flights in regions where detail is specially important. The final maps would not, of course, consist of photographic pictures, but would be conventional drawn maps; the photographs of each area would, however, always be available, and presumably prints could be supplied at any time to people particularly interested in any special area, in order that detail not included in the conventional map could be studied. A study of the foreground of the photograph facing p. 90, and of the stereoscopic view carried in a pocket in the cover, will give some idea of the value of such photographs to anyone contemplating the development of any area.

It may happen, however, that some areas are of very much more interest than others and of these, if they are suitable, mosaic maps might be constructed, either during the main survey, or at some subsequent date; where this is done the reconnaissance map would provide the ideal framework upon which to carry out the flying for the mosaic map. Some such framework would have to be made in any case before the flying for the mosaic could be started.

Other areas may be important enough to warrant something more detailed and accurate than the reconnaissance map, but they may not be suitable for mosaic work, either because they are too hilly, or because accurate contours of height are required. For these areas the original oblique photographs might be re-sected by reference to a sufficiently close ground control, which could be laid down at any subsequent time when the increasing importance of the area warrants the extra expense. For this to be possible it would merely be necessary for the original photographs to have been taken in cameras which are, optically, rather more accurate than would be strictly necessary for the reconnaissance work. The provision for this greater accuracy in the cameras would involve only a negligible increase in the first cost of the survey. To be economically practicable, this re-section would require the use of elaborate specialised apparatus, such as the Auto-Cartograph.

We may state the matter more generally by saying that the methods of navigational control, coupled with oblique photographs, would allow the whole of an unknown area to be systematically photographed without any co-operation from the ground; and the photographs so obtained could, if desired, be used to construct a rough but continuous reconnaissance map of the area, without any help from ground control. The map, if made without help from ground surveying, would not be very accurate, because it would contain the full errors involved in aerial navigation by compass and airspeed indicator, but the provision of a very open ground control, in which the control points were spaced as far apart as 50 miles, or even more, would enable the map to be adjusted with the accuracies stated in this chapter and in chapter VI. The accuracy of this map would depend upon the closeness of the available ground control, and it could be progressively increased from time to time as more ground control becomes available. Ultimately, when the ground control becomes sufficiently close, the methods of re-section could be applied to give a map with accurate height contours. The reconnaissance map, in any of its stages of progressively increasing accuracy, forms the ideal frame upon which to build the more expensive but more detailed mosaic map, which could be added in important and suitable areas, either during the original survey, or at a later date. Finally, it is to be noted that the photographs taken in the original aerial survey could be used again and again throughout the progressive improvements in the accuracy of the map; until, in fact, the development of the country has so altered its features that a new survey is required.

THE METHODS USED IN THE STUDY OF
RANDOM ERRORS

THE causes of error in the observation of a quantity, or in the adjustment of an apparatus, may be divided into two classes,."Systematic" and "Random." The mean of a large number of observations upon a given quantity will have an error approximating to that due to the systematic causes, for the errors due to random causes will, in the long run, neutralise each other. The random errors of a series of observations may therefore be looked upon as the deviations of individual errors from the mean error of a large number of observations of one type.

Although random errors might conceivably have any magnitude, yet it is obvious that the probability of the occurrence of a large error will be greater in a series of inaccurate observations than in an accurate series. If, therefore, we wish to study the accuracy of a series of observations, we must agree upon some method of defining accuracy in terms of the probability of the occurrence of any given error. One method of doing this is to state the probabilities that the error will be less than some assigned magnitudes. Thus, we may say that 10 per cent. of the errors will be less than some magnitude, 20 per cent. less than some other magnitude, and so on. This information could either be given in a table, or set out in the form of curves, such as those in Diagrams 12 to 15.

It is shown in books on the theory of probability that, for "scalar" quantities —that is quantities that can be defined by a single variable, for example, the height of a point above the horizontal plane—random errors will tend to be distributed so that this curve will have a certain definite shape. This remarkable conclusion is only strictly accurate when the term "random" error is defined in a very precise manner, which we need not discuss, but it is found in practice that most errors which would normally be defined as random are distributed, when enough are taken, in approximately this manner. This particular distribution is called the "normal" distribution of errors.

The degree to which a set of errors are distributed in accordance with the "normal" distribution, appropriate to purely random errors, can be ascertained by arranging the observed errors in order of magnitude and plotting them, as in Diagram 15, where the errors appear as ordinates and the points are equally spaced in the direction of the abscissae. The resulting chain of points will also show the percentage of the total number of errors which have less or more than any assigned magnitude. A curve corresponding to "normal" distribution can then be drawn on a scale which will cause it to lie as evenly as possible between the observed points, and the degree to which the distribution of the observed errors approximates to the "normal" distribution will then be shown by the extent to which the observed points follow this curve throughout its length. Diagram 14 shows two sets of observations for which the distribution is very close to the normal, whereas in Diagram 15 it was impossible to obtain a close fit between the normal distribution curve and the chain of observed points, thus showing that the distribution was not quite normal.

Before any exact application can be made of the theorems which follow from the assumption that the error distribution in a set of observations is normal, it is necessary, first, to ascertain by reference to a very large number of observations whether this really is the case. When, however, it is only desired to obtain a rough idea of the order of accuracy of certain observations, it is possible, with a little experience, to distinguish at sight those types of error which can, for this purpose be classed as "random."

Once it is known that the distribution of errors in a set of observations is sufficiently nearly normal for the purpose in hand, the whole information relating to the accuracy of the observations can be given by stating the magnitude of the error which will have any given chance of exceeding the observed error. For many reasons it is convenient to choose for this purpose the error which has a 50 per cent. chance of exceeding the observed error. This error is used throughout the present work to define the accuracy of any set of observations; it is called, simply, the 50 per cent. error*. Similarly the error that has any other chance of exceeding the observed error, say a 90 per cent. chance, is named simply by the value of the chance, as the 90 per cent. error. The 50 per cent. error is, however, the one which we have generally used for the purpose of defining accuracy; others are only included in special cases.

When the quantity under consideration is not "scalar" but is in the form of a vector depending upon two components, the curves of error distribution will not be the same as for scalar quantities; their form will depend upon the ratios of the probabilities of error in the two components. When, in the present work, we have to deal with vector quantities—such as the tilt of the camera axis from the vertical, or the vector change in the wind—it happens that the components of the vectors are subject to about equal chances of error, and in these cases the distribution curve takes the form shown in Diagram 12.

For rapid estimation the following figures have been extracted from the normal distribution curves for scalar and vector quantities. For convenience, the 50 per cent. errors have been taken as unity.

Nature of observed quantity	Scalar	Vector with two equally accurate components
50 per cent. error	1·0	1·0
90 ,, ,,	2·4	1·8
99 ,, ,,	3·8	2·6
99·9 ,, ,,	4·9	3·2

Using this table, we can, given the 50 per cent. error of a set of observations, estimate the errors that may be expected to occur once in every ten, a hundred, or a thousand, observations; provided always that the distribution is strictly normal. Given the 50 per cent. error one can, in practice, generally rely upon the 90 per cent. error being approximately as indicated in the table in the case of any set of errors which one would reasonably describe as random, but before much reliance can be placed upon the exact value of the 99 per cent. and 99·9 per cent. errors, the distribution curve for the type would have to be investigated very thoroughly, by reference to a large number of observations.

* This error is often called the "Probable" error, but many people object to this term because it implies that it is the one which is most likely to occur, and this is not so.

We may describe these results in a convenient practical way by saying that errors greater than twice the 50 per cent. error must be expected to occur occasionally, but that errors greater than three times the 50 per cent. error should be very rare, particularly in the case of vector errors.

When the distribution of a set of errors is normal, certain very useful theorems can be deduced. We shall state two of these without the proofs, which can be found in any book on the theory of errors.

(1) When the errors are due to a number of *independent* causes, the first of which, acting alone, would lead to a 50 per cent. error of say a_1, the second to an error a_2, and so on, then the 50 per cent. error of a set of observations in which all the causes act together would be $\sqrt{a_1{}^2 + a_2{}^2 + a_3{}^2 + a_4{}^2 + \text{etc.}}$

(2) The means of a number—n—of independent observations of the same kind will have a 50 per cent. error equal to $1/\sqrt{n}$ times the 50 per cent. error of the individual observations.

These rules are of great practical importance and are often used in the present work. A short experience of experimental work involving the study of errors leads one to apply them with great confidence in all cases where an approximate figure only is required to represent the accuracy of a set of observations. Care must, however, be taken to exclude systematic errors, and to avoid applying the results to simultaneously acting causes of error which are not mutually independent.

When a finite number of observations is being studied, an approximate value for the 50 per cent. error can be very quickly found by dividing the errors into two equal groups, such that all the errors in one are greater than any error in the other; the approximate 50 per cent. error can then be taken to lie between the largest error of the smaller group and the smallest error of the larger group. A more accurate method is to plot the errors, as in Diagrams 12 to 15, and, after fitting the normal distribution curve to them, record the 50 per cent. error of this curve. It requires only a surprisingly few observations, treated in this way, to determine the 50 per cent. error of the type, with sufficient accuracy for most practical purposes.

APPENDIX II

PILOTS REPORTS OF THE FLIGHTS DISCUSSED IN CHAPTER IV

THESE REPORTS WERE ALL RECEIVED BEFORE THE FLIGHTS HAD BEEN ANALYSED

Report by Fl.-Lt. Coleman

Diagram 1. *Date* 24. ii. 21.

Thirty-six plates were taken up and on taking the first eighteen it was found that they did not pass into the empty box after exposure, so we changed boxes and commenced again flying from the N.E. end of the canals to the S.W. end.

On the first journey the weather was very good but on the return* it was somewhat bumpy, and I should think that in the first two or three plates taken there will probably be a good deal of tilt, but the remaining fifteen should be fairly good.

I used the method of flying straight of flying with a shadow cast by the sun on the dashboard in the same place.

Report by Flying Officer Charles Allan

Diagrams 2, 3. *Date* 1. iv. 21.

Approached the Canals about 5 miles distant in direct line of same to get drift correction before actually taking the photographs. Then kept straight and level by flying on distant point and horizon, glancing down through bomb sight occasionally to see if still over the Canals. Turned and did same coming back. Height nearly constant. Speed 83 m.p.h. average. Air very calm.

Report by Fl.-Lt. Coleman

Diagrams 4, 5, 6. *Date* 24. vi. 21.

We left the ground at 10.30 a.m. and started with the photographs at about 11 a.m.

The first box† we took of the Bedford Canals flying from the S.W. to N.E. end. I trued up the machine to fly at 80 m.p.h. indicated air-speed and got my drift before giving Griffiths the signal to commence. The machine was flying slightly left wing down and, owing to the stiffness of the throttle control, experienced a little difficulty in keeping the machine at a true fore-and-aft level.

The second box we started from the N.E. end and flew to the S.W.; the trip was very much the same as the first.

The third box however, when returning again from the S.W. to the N.E. end, should be better, as I managed to get the correct position of the throttle lever and the machine appeared to fly more level fore-and-aft and this I managed to keep constant.

During all the trips we encountered small bumps laterally but this should not affect the tilt much.

In every trip I flew straight by the shadow method.

---

* It is the return journey that is recorded in Diagram 1.
† The pilot is referring to his box of 18 plates.

Report by Fl.-Lt. Coleman

Diagrams 7, 8, 9. *Date* 25. vi. 21.

7. Course N.E. to S.W. Conditions were good and the results should be good as the machine appeared to fly very steadily. A slight tilt to the left might be evident owing to the machine being slightly left wing low. I swung the machine slightly for the first few plates, but I do not think it was enough to affect them. The height was constant.

8. Course N.E. to S.W. Conditions as above. In my opinion this box should be the best as I managed to keep the machine steadily on its course without swing or loss of height.

9. Course S.W. to N.E. Conditions as above. I had trouble in getting the machine on my course and the first four or five plates will probably show a good deal of tilt both laterally and fore-and-aft. The remaining plates however should be good as I overcame the difficulty and kept a very steady course and constant height.

In all my flights I used my instruments together with a shadow cast by the sun for flying straight. This particular machine D.H. 9 A J. 567 swings slightly, probably due to slackness on the rudder controls. These will be tightened.

Report by Flying Officer Charles Allan

Diagram 10. *Date* 28. ix. 21.

Course S.W. to N.E. Experienced difficulty in keeping straight. No object handy on which to fly. The conditions were fair with thick haze. Indicated airspeed 83 m.p.h.

Report by Fl.-Lt. Coleman

Diagram 11. *Date* 5. x. 21.

Using gyro rudder. control.

Course from N.E. to S.W. Conditions fair, very misty, no clouds. This box should turn out excellent as I got the drift correct and was able to use the gyro without interruption the whole length. This should give us the information we require as to the accuracy of the rudder control. In this last box, I noticed that there was not the slightest loss in height as shown on the pilot's aneroid, and the reading given by the observer from his show no loss or gain. I also noticed that the bubble statoscope* remained in a level position. Height 10,200 ft.

---

* This instrument which is a rate of climb recorder was carried at the time, but was apparently not much used and later it was not carried.

PART I. ON THE INFLUENCE OF A STEADY RATE OF CHANGE OF WIND UPON THE ACCURACY OF NAVIGATIONAL CONTROL

Let an aeroplane starting from A fly in such a way that, if the wind did not change, it would reach B in time t_0.

Let the wind acting upon the aeroplane be changing in such a way that, after any time t, the vector change is at, in a direction parallel to BC.

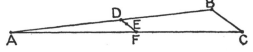

On account of this change the aeroplane, after time t_0, will be at C instead of at B, where $BC = \frac{1}{2}at_0^2$.

At any intermediate time, t, it would, in the absence of the wind change, have been at D on AB such that $AD/AB = t/t_0$, but, owing to the wind change, it will actually be at E, where DE is parallel to BC and equal to $\frac{1}{2}at^2$.

The system of navigational control, working from A and C as known points, would indicate the position of the aeroplane, after time t, to be at F on AC, such that $AF/AC = t/t_0 = AD/AB$.

DF is therefore parallel to BC and E lies on DF. EF is the error of the navigational control in this case and we have

$$EF = DF - DE = BC \cdot t/t_0 - \frac{1}{2}at^2.$$

Let $\qquad AF = x \cdot AC$ so that $x = t/t_0$.

Then $\qquad EF = BC \cdot x - \frac{1}{2}at_0^2x^2 = \frac{1}{2}at_0^2 \cdot x\,(1 - x)$.

Let the vector change of wind during the flight be v, so that $v = at$.

Then $\qquad EF = \frac{1}{2}vt_0 \cdot x\,(1 - x)$ ....(1).

This error has a maximum value, for given values of v and t_0, when $x = \frac{1}{2}$, that is half-way through the flight by time, and this maximum value is

$$\tfrac{1}{8}vt_0 \qquad ....(2).$$

We may put this result into words by the statement that the maximum error due to a uniform rate of change of wind is one-eighth the product of the vector change of wind during the flight and the time of flight.

In the *Aeronautical Journal* for May 1921, p. 226 an expression for the 50 per cent. vector change* of the upper wind with space and time, in the British Isles, is given by Dobson in the form

$$\text{50 per cent. change with space} \quad \ldots \quad 0\cdot5\,\sqrt{d},$$
$$\text{50 per cent. change with time} \quad \ldots \quad 3\cdot3\,\sqrt{t},$$

where d is distance in miles and t time in hours.

The square roots in these formulae cannot of course represent the actual facts exactly, since they would lead to infinite changes of wind after the lapse of an infinite time, but provided that they are only applied over moderate times and distances, say a few hours and not more than 100 miles, they do appear to give a

---

* By this is meant the vector change which is equally likely to exceed, or be exceeded by, an individual vector change. See Appendix I.

moderate approximation to the observed facts. The former of the above formulae receives some support from data collected by Durward* from certain meteorological stations in France, although these data lead to a slightly higher value for the coefficient, about 0·65 as against the 0·5 given by Dobson. We will however take Dobson's figures as the basis of a rough estimate of errors.

If we may assume that the change of wind with time is independent of its change with space, we may express the 50 per cent. vector change of wind during a flight of length d and duration t in the form

$$\sqrt{0\cdot5^2 d + 3\cdot3^2 t}.$$

Using equation (1) the 50 per cent. vector error in the middle of a flight can therefore be expressed in the form

$$\tfrac{1}{8} t \sqrt{0\cdot25 d + 10\cdot9 t}.$$

It is true that this expression assumes a steady rate of change of wind on the aeroplane, which is not consistent with Dobson's formulae, but the error introduced in this way is small.

Let us assume that the ground speed of the aeroplane is in the neighbourhood of 80 miles per hour. The 50 per cent. vector error then becomes

$$\tfrac{1}{8} \cdot \tfrac{1}{80} \sqrt{0\cdot25 + 0\cdot13\, d} \sqrt{d},$$

which, when expressed as a percentage of the length "d" of the flight, reduces approximately to $0\cdot1 \sqrt{d}$.

Thus, in a flight of 50 miles we should, on these data, expect the error near the middle to exceed 0·7 per cent. of 50 miles, or 0·35 mile, on every other occasion, whilst on a flight of 100 miles these figures would be increased to 1 per cent. or 1·0 mile.

PART II. ON THE ELASTIC ADJUSTMENT OF ERRORS IN A NAVIGATIONALLY CONTROLLED SURVEY

Let the actual length of the flight AC be given the symbol d and the average ground speed during the flight the symbol V, then equation (1) part 1 becomes

$$EF = \frac{1}{2} \frac{v}{V} d \cdot x (1 - x) \dots \dots \dots (3).$$

Consider a uniform elastic string, stretched between two points A, C distant l apart, and let the ratio of the tension in the string to its linear strain be E.

Consider a point F on the string such that $AF = x \cdot d$.

If this point is displaced by a distance ϵ perpendicular to the string, the restoring force is

$$T \left[\frac{\epsilon}{xd} + \frac{\epsilon}{d - xd} \right] = \frac{T\epsilon}{d} \frac{1}{x (1 - x)}.$$

If the same point is displaced by a distance ϵ along AC, the restoring force is

$$E \left[\frac{\epsilon}{xd} - \frac{\epsilon}{d - xd} \right] = \frac{E\epsilon}{d} \frac{1}{x (1 - x)}.$$

---

* An Air Ministry Publication, "Professional Notes," No. 24.

Under the action of a deflecting force having components P and Q perpendicular and parallel to the string the deflections of the point F would therefore be

$$\frac{Pd}{T} x (1 - x) \text{ and } \frac{Qd}{E} x (1 - x) \dots\dots\dots\dots(4).$$

Comparing these expressions with equation (3) we see that, in both cases, the deflection that results from a given load on the string varies, along the string, in the same way that the error involved in the method of navigational control varies along the flight, provided that this error is due solely to a uniform rate of change in the wind acting on the aeroplane.

If, as we saw in chapter VI, the probability of error is nearly the same in directions parallel to and perpendicular to the direction of the flight, and if the probable vector change of wind during a flight is the same for all flights, it follows that a net of elastic strings for which $E = T$, stretched between the known points at the end of each flight and treated as in chapter VI, p. 85, will give a good approximation to the ideal error adjustment that is there required; for every point of every string will be held in position by its own string, with a rigidity inversely proportional to the probability of error of its location in the corresponding flight; hence each flight, at each point, will play its due part in the general adjustment.

Actually the probability of the vector change "v" exceeding any given magnitude will depend to some extent on the length and time of the flight; so that, to be strictly accurate, longer flights should be represented by weaker strings; if, however, the variation of length in the flights is not very great, it is probably not worth attempting to apply this refinement.

It may be noted that, should it be found that the probability of error in the direction of a flight is different from that perpendicular to the flight, the ratio of T to E in the elastic strings could be correspondingly varied.

If a material obeys Hook's law, T is equal to E when the material is stretched to twice its unstrained length. Rubber bands do not exactly obey this law and difficulty was experienced in making $T = E$. The inequality was slight and was ignored, the strings being simply stretched to twice their natural length. The relation of T to E in any state of strain is easily tested as follows. Stretch a length of material between two points on a board and hang a weight near the middle. Hold the board so that the string is first horizontal and then vertical, and in each case note the displacement of the weight.

ON THE CONSTRUCTION OF GRATICULES FOR THE MEASUREMENT OF RELATIVE AZIMUTH ANGLES AND DEPRESSIONS BELOW THE HORIZON

The equi-azimuth lines in this graticule will be the intersections of the photographic plane with a series of vertical planes passing through the optical centre of the lens; they thus consist of a pencil of straight lines radiating from the plumb point of the photograph. The equi-depression lines are the intersections with the photographic plane of a series of cones, having a common vertical axis and their apex at the optical centre of the lens; they thus consist of a family of conic sections, ranging from a straight line representing the horizon, through a series of hyperbolae and ellipses, to a point at the plumb point on the photograph. Any one graticule will not, of course, contain the whole of this series, owing to the limited field of view of a photograph, and the series will differ in form according to the tilt of the optical axis of the camera from the vertical.

The graticules could theoretically be constructed by photographing, from appropriate angles, a series of radiating straight lines and concentric circles, and it is probable that this would be the simplest method to adopt in practice when the optical axis is only slightly inclined to the vertical; but when the photographs are very oblique, and extend perhaps to the horizon, this method would not be practicable, and it becomes necessary to draw out the graticule by calculation, or by mechanical means. It will generally be advisable to draw out the graticule on a large scale and reduce it photographically, and when this is done, it may be practicable, by photographing the drawing from different angles, to make one drawing suffice for a series of graticules suitable for different camera tilts. It is to be noted that if the graticule is photographically reduced in the same camera which is used for the aerial photography, errors due to aberration of the lens will be avoided.

There are many different ways in which such a graticule could be drawn; in the following notes the method which was used to construct the graticule facing p. 90 is described, and the errors involved in the use of a graticule suitable for a tilt differing slightly from that of the photograph are discussed.

In fig. 5 the plane of the paper is a vertical plane containing the optical axis of the camera. The optical centre of the lens, O, is however shown on the wrong side of the photographic plane. This is done for convenience, and is merely equivalent to studying the print instead of the negative. In fig. 4 the plane of the paper is the plane of the photograph.

A is the optical centre of the photograph and AH is perpendicular to the horizon trace on the photograph, so that A and H are identical points in the two figures. AH is described as the zero azimuth line, and the azimuth of other points is measured from this line. ph is another azimuth line, radiating from the plumb point; the co-ordinates of p with reference to HA and Hh are x and y, and are expressed as multiples of the focal length f, which is itself equal to OA in fig. 5.

Let θ be the depression of the optical axis below the horizon.

Then the distance, on the photographic plane, of the plumb point V from the horizon trace is HV, where

$$HV = OH/\sin\theta = f/\cos\theta\sin\theta = f \cdot 2\operatorname{cosec} 2\theta \ldots\ldots\ldots\ldots\ldots(1).$$

Let ϕ be the azimuth of the line *ph*.

Then $\qquad\qquad Hh = OH \tan \phi = f \sec \theta \tan \phi \qquad$ ....(2).

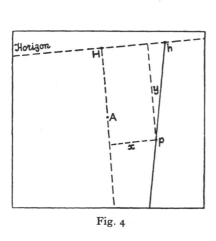

Fig. 4

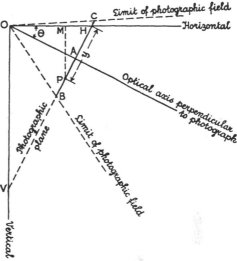

Fig. 5

The pencil of equi-azimuth lines radiates from a point on *HA* produced distant $f . 2 \operatorname{cosec} 2\theta$ below the horizon and intersects the horizon at points distant $f \sec \theta \tan \phi$ from *H*. These lines are thus easily drawn.

The equi-depression lines are more difficult to draw.

Let ϵ be the depression below the horizon of any point of which the co-ordinates are x, y.

When the point is on the line *HA*, so that x is zero, the distance of the apex of the hyperbola below the horizon trace is

$$f\,[\tan \theta - \tan (\theta - \epsilon)] \qquad ....(3).$$

The apices of the hyperbolae are therefore easily found.

When x is not zero we have

$$\tan \epsilon = PM/(OM^2 + f^2 x^2)^{\frac{1}{2}},$$

where $PM = fy \cos \theta$ and $OM = f\,(\sec \theta - y \sin \theta)$.

Hence $\qquad\qquad x^2 = y^2 \cos^2 \theta \cot^2 \epsilon - (\sec \theta - y \sin \theta)^2 \qquad$ ....(4).

The depression curves could be computed from this equation, but the computation would be difficult, especially for small values of x. An alternative procedure, which was adopted in the construction of the graticule facing p 90, is as follows.

Regard the above equations as a quadratic for y, solve it, expand the result in ascending powers of x, and transfer the origin of co-ordinates to the apex of the hyperbola. This leads to the equation

$$y = \tfrac{1}{2}x^2 \tan \epsilon - \tfrac{1}{8}x^4\,[1 - \tan^2\theta \tan^2\epsilon]\,\cos^2\theta \tan \epsilon + \text{etc.} \qquad ....(5).$$

Compare this with the equation for the circle which osculates with the hyperbola at the apex and has a radius equal to $f \cot \epsilon$

$$y = \tfrac{1}{2}x^2 \tan \epsilon + \tfrac{1}{8}x^4 \tan^3\epsilon + \text{etc.} ....(6)$$

The distance between y co-ordinates for these two curves is

$$\tfrac{1}{8}x^4 \tan \epsilon \, [(1 - \tan^2\theta \tan^2\epsilon) \cos^2\theta + \tan^2\epsilon]$$

or
$$\tfrac{1}{8}x^4 \tan \epsilon \sec^2\epsilon \cos^2\theta \qquad \dots\dots\dots\dots\dots\dots(7).$$

This difference reaches its maximum value in the bottom corners of the photograph where it amounts, in the illustration facing p. 90, to a little over one-hundredth of an inch, or roughly one-tenth of a degree.

In the case of this graticule, therefore, the equi-depression lines could be made with sufficient accuracy as follows. Mark off, on the line AH, the apices of the hyperbolae, at points distant $f\,[\tan \theta - \tan (\theta - \epsilon)]$ below the horizon. Set a beam compass to a radius $f \cot \epsilon$ and, with its centre on AH produced, describe a circle passing through the appropriate apex. It will be found that lines corresponding to depressions less than some 6 degrees will involve inconveniently large compasses, but these, being nearly straight, can be easily drawn by plotting a few points from equation (5) and joining by straight lines.

If photographs covering a wider angular field are to be used, the errors in the above approximation become too large and a second approximation is required, this can be obtained by slightly increasing the radius of the circles, so that instead of lying wholly below the hyperbolae they lie above them for small values of x and below them for large values. It can be shown that the best approximation is obtained in this way when the radius is

$$f \cot \epsilon/(1 - 0{\cdot}208 \cos^2\theta \sec^2\epsilon \,.\, x_0{}^2),$$

where x_0 is the greatest value of x for which the graticule is required.

Using this corrected radius the maximum error is reduced to about one-sixth of the maximum error when the radius is equal to $f \cot \epsilon$. This expression for the radius was used in the construction of the graticule facing p. 90.

Finally let us make an approximate estimate of the errors involved by using a graticule suitable for a tilt slightly different from the true tilt.

The equation of the equi-azimuth lines is

$$x = (\sec \theta - y \sin \theta) \tan \phi.$$

Let the tilt for which the graticule is used differ from the true tilt by an angle $\delta\theta$. Any point for which the azimuth is ϕ will be displaced from its correct position, in the direction of the axis of x, by an amount

$$\delta x = \frac{\partial x}{\partial \theta}\, \delta\theta.$$

But
$$\frac{\partial x}{\partial \theta} = \left(\frac{\sin \theta}{\cos^2\theta} - y \cos \theta\right) \tan \phi.$$

The displacement for a given ϕ will be greatest when $y = 0$ and will amount to

$$\frac{\sin \theta \tan \phi}{\cos^2\theta}\, \delta\theta.$$

In the graticule facing p. 90, θ was $17°$ and the greatest value of ϕ on the horizon line was $20°$, hence the greatest error involved in this case is approximately $0{\cdot}1\, f\delta\theta$. If graticules are provided for every whole degree, as suggested in chapter VI, and if $f = 6$ inches, this error amounts to $0{\cdot}005$ inch or corresponding roughly to $0{\cdot}05$ degree.

The approximate equation for the depression lines relative to the horizon as axis of X is

$$y = \tan \theta - \tan (\theta - \epsilon) - \tfrac{1}{2}x^2 \tan \epsilon;$$

whence
$$\frac{\partial y}{\partial \theta} = \sec^2\theta - \sec^2(\theta - \epsilon),$$

and the error involved by using a wrong graticule is approximately
$$\delta y = \frac{\partial y}{\partial \theta}\,\delta\theta = (\sec^2\theta - \sec^2[\theta - \epsilon])\,\delta\theta.$$

This error has, in the case where the horizon is near the top of the photograph, a maximum value at the optical centre where $\epsilon = \theta$; in which case
$$\delta y = (\sec^2\theta - 1)\,\delta\theta.$$

When $\theta = 17°$ this becomes $\delta y = 0.09\,\delta\theta$, and the maximum error $= 0.09\,f\delta\theta$, which, when $f = 6$ inches and $\delta\theta = \frac{1}{2}$ degree, is equal approximately to 0.005 inch or 0.05 degree.

When therefore an accuracy of 0.1 degree is all that is required, it is amply sufficient to provide graticules for every whole degree and to use that graticule which is most nearly suited to the actual tilt of the photograph.

ON THE PROBABILITY OF ERROR IN THE ESTIMATION OF HEIGHT BY THE METHODS OF CHAPTER VI

The height of any point may be expressed in the form

$$H - L \tan E,$$

where H is the height of the aeroplane;

L is the horizontal distance from the plumb point;

E is the depression below the horizontal plane of the point as seen from the aeroplane.

If errors δH, δL and δE occur in the estimated values of H, L and E, the resulting error in the estimated height of the point will be

$$\delta H - \delta L \tan E - \delta E L \sec^2 E,$$

which may to a first approximation be written

$$\delta H - H_1 \partial L / L - \delta E L \sec^2 E,$$

where H_1 is the height of the aeroplane above the point.

The errors involved in these three terms are entirely independent of one another; hence the 50 per cent. error of a single observation of the height of a point can be written

$$\sqrt{h^2 + (H_1 l/L)^2 + (eL \sec^2 E)^2},$$

where h, l and e are the 50 per cent. errors of H, L and E.

Consider now a point near the middle of one of the squares made by the flights, when using the methods of chapter VI; it will appear in from 8 to 12 photographs, taken from four directions approximately at right angles to each other. Each of these photographs will allow a separate determination of height to be made, so that the final error will be reduced by taking a mean of all the individual observations. The reduction of error cannot, however, be assumed to be in the ratio of $\sqrt{n}$ to 1, where n is the number of photographs concerned, because some of the sources of error are systematic to several photographs. Let us consider the three sources of error separately.

The error h will be dependent upon the existence of a temperature gradient different from that assumed when correcting the aneroid, and upon errors in the aneroid itself; these causes may well persist for several days, so that, to be on the safe side, we shall assume that h will not be reduced at all by taking means.

The error $H_1 l/L$ results from errors in the assumed scale of the map, along the line in question. In those cases where the map is constructed upon a close ground control, such errors will be small and not systematic, so that they will produce little influence on the final error after taking means. In cases where the ground control is very widely spaced, on the other hand, errors in scale may occur which will seriously influence the estimation of height. If we consider the method discussed in chapter VI, for the construction of aerial maps, we realise that systematic errors of scale in any given direction are to be expected, but that errors of scale in directions which are mutually perpendicular should be quite independent of each other. Now, by this method, observations relating to height will be taken along rays which can approximately be divided into two groups which are mutually per-pendicular, hence the process of meaning should reduce the errors due to this cause in the ratio $\sqrt{2}$ to 1.

Errors depending upon e are due partly to random errors in the identification of the point and the estimation of its position on the photograph, and partly to inaccurate knowledge of tilt. We saw that the 50 per cent. error due to the first two causes combined can be kept down to 0·1 degree even when working by very rapid methods, and this cause of error being entirely unsystematic, will there-fore have little influence on the final result. The other part of the error ϵ due to the tilt not being accurately known cannot, however, be considered as entirely unsystematic, for, in cases where the error is due to errors in piloting, the pilot may, from our experience, be flying with one wing permanently down, and in cases where the tilt is referred to the horizon, the same horizon will be used in all the photographs of one flight. In both these cases, however, the errors of tilt estimation are not systematic from flight to flight and, since there are four flights concerned, we may safely assume that the 50 per cent. error due to this cause will be reduced, by taking means, in the ratio 2 to 1.

From the above considerations we may write the probable magnitude of the final error in the estimation of the height of a point near the middle of the squares in the form

$$\sqrt{h + (1/\sqrt{2}\, H_1 l/L)^2 + (1/2\, L \sec^2 Ee)^2}.$$

Let us evaluate this expression in a particular case, assuming

$$H_1 = 10{,}000 \text{ ft. and } L = 5 \text{ miles or } 26{,}400 \text{ ft.}$$

If the aneroid height is recorded in the film and corrected for temperature, as described in chapter VII, h should not exceed 50 ft.

l/L will depend upon the ground control employed and is difficult to estimate. We may for example take the experimental results of Diagram 17, which are not likely to be better than will be obtained by practised pilots on routine work; here the 50 per cent. error in the estimated length of a side of any of the squares into which the flights divide the area comes out well under 1 per cent. of the length of the sides. If, to be on the safe side, we call this error 1 per cent. so that $l/L = 0\cdot01$, we find that $1/\sqrt{2}\, H_1 l/L - 71$ ft.

In the last term $E = 20°$ approximately, hence $\sec^2 E = 1\cdot14$.

e we have seen in chapters III, IV and VI will be about 1 degree when no horizon is available and 0·2 degree when the horizon is used. Hence we have

$$1/2\, L \sec^2 Ee = 260 \text{ ft. without horizon,}$$
$$= 52 \text{ ft. with horizon.}$$

Inserting these figures into our expression for the final 50 per cent. error we get the following results:

Without horizon ... ... 274 ft.
With horizon but without close ground control ... ... 100 ft.
With horizon and with ground control sufficient to prevent
systematic errors of scale ... ... 72 ft.

These figures refer to points that occur near the middles of the squares formed by the navigationally controlled flights, when the sides of the squares are some ten miles in length. Points not near the middles, and not so closely under the flights that $\sec^2 E$ becomes large, will be more accurately determined as to height by the nearer photographs and less accurately by the more distant ones, so that it seems probable that the height of all points which do not lie within a mile or so of any flight will be determined with about equal accuracy. The points for which it will be most difficult to estimate height correctly will be those which occur close beneath the intersection points of two flights, for in these cases reliance will have to be placed upon photographs taken from a distance of some ten miles. The error due to *e* will in these cases be doubled and the probable error of height-estimation after taking means would rise on this account to the following magnitudes:

Without horizon ...   .   530 ft.
With horizon but without close ground control   133 ft.
With horizon and with a close ground control   112 ft.

These increased errors refer only to the relatively small areas that lie closely under the aeroplane tracks.

From the foregoing considerations it is reasonable to draw the following general conclusions relating to the methods of surveying described in chapter VI:

(1) If we have to rely entirely on good flying for our knowledge of camera tilt, it will not in general be practicable to determine absolute height with any useful accuracy.

(2) When the camera tilt is determined by reference to the horizon the probable error of height determination should be in the neighbourhood of 100 ft. Occasional errors greater than 200 ft. are therefore to be expected.

APPENDIX VI

ON THE CORRECTION OF THE AIR-SPEED INDICATOR TO ALLOW FOR VARIATIONS IN AIR DENSITY

In chapter VII we discussed methods of calibrating the air-speed indicator for air of some particular density, but noted that, when great accuracy is required, a separate correcting factor can be found only after measurements have been made of both the temperature and pressure of the air around the aeroplane. We have to consider methods by which this correcting factor can be found and conveniently applied by the observer.

Let V_i be the reading of the instrument,

V_p* be the corresponding reading of a perfect instrument of the pitôt type,

V_a be the true air-speed;

then
$$V_a = [\rho_0/\rho]^{\frac{1}{2}} V_p = K [\rho_0/\rho]^{\frac{1}{2}} V_i \quad \dots\dots\dots\dots\dots\dots(1),$$

where ρ_0 stands for "standard density" for which the instrument is calibrated;

ρ stands for the density of the air at the height of the aeroplane;

K is the factor, determined in the calibration of the instrument, which makes the readings correct in air of standard density.

We have to determine the quantity $[\rho_0/\rho]^{\frac{1}{2}}$ and we do this from the well-known relation

$$\frac{\rho_0}{\rho} = \frac{p_0}{p} \cdot \frac{\theta}{\theta_0},$$

where p and θ are the pressure and temperature (absolute) corresponding to the density ρ, whilst p_0 and θ_0 correspond to ρ_0.

Now the altimeter is incidently a pressure indicator, although it is not graduated directly in terms of pressure, but in terms of the equivalent height according to an arbitrary law. It is not therefore necessary to carry a separate pressure indicator, provided that the readings of the altimeter can be correctly interpreted in terms of pressure.

British service altimeters are graduated according to the law

$$h' = - A \log_{10} p/p_0,$$

where A is a constant equal to 62,582, p_0 is the pressure due to 760 millimetres of mercury, and h' is the height written against the position of the pointer when the pressure is p.

We may re-write this equation in the form

$$p/p_0 = 10^{-h'/A}.$$

Now the scale of heights in the altimeter can be moved, as a whole, to allow the zero to come opposite the needle at the height of the ground from which the flight starts. Let the shift of the scale, from the true zero corresponding to the pressure p_0, be h_g. Then, provided that the scale of heights is uniform, as is the case in the standard altimeters, the true reading from the point where the pressure is p_0 will be $(h + h_g)$, where h is the actual reading on the shifted scale.

---

* V_p is sometimes called the "indicated air-speed" and sometimes the "pitôt speed."

Hence $\qquad h_1 = h + h_g$

and $\qquad\qquad p/p_0 = 10^{-(h+h_g)/A}$

$$= 10^{-h_g/A} \times 10^{-h/A}.$$

But $10^{-h_g/A}$ is the ratio of the pressure at the starting-point to the pressure p_0. Let us call this ratio b/p_0, where b stands for the barometric pressure on the aerodrome.

Hence $\qquad p/p_0 = b/p_0 \times 10^{-h/A}.$

But $\qquad\qquad \dfrac{p_0}{\rho} = \dfrac{p_0}{p} \cdot \dfrac{\theta}{\theta_0};$

therefore $\qquad \dfrac{p_0}{\rho} = p_0/b \times 10^{h/A} \times \dfrac{\theta}{\theta_0}.$

But $\qquad\qquad V_a = K\,[\rho_0/\rho]^{\frac{1}{2}} \times V_i;$

therefore $\qquad V_a = K\,[p_0/b]^{\frac{1}{2}} \times \left(10^{h/A} \times \dfrac{\theta}{\theta_0}\right)^{\frac{1}{2}} \times V_i.$

This expression allows the true speed to be found in terms of the readings of the speedometer. In aerial surveying we do not however require to know the speed explicitly, but rather the time required to cover some given distance on the ground, and for this purpose we make use of the above result in the manner described in Appendix VII.

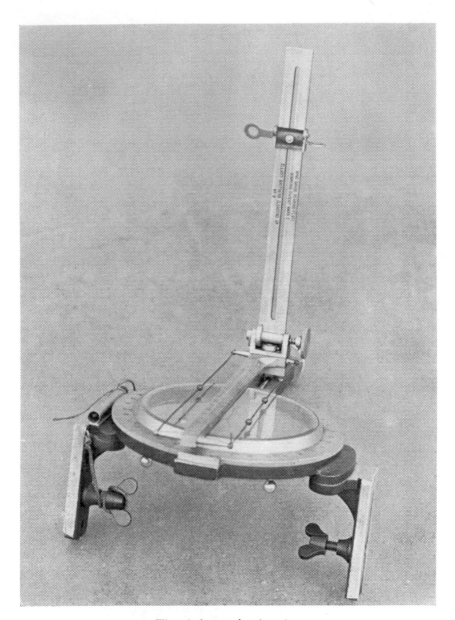

The wind gauge bearing plate

APPENDIX VII

ON THE WIND GAUGE BEARING PLATE AND ITS USE
IN AERIAL SURVEYING

This instrument, illustrated opposite, is used to find the strength and direction of the wind at the survey height, and to calculate the direction in which to fly and the time interval between photographic exposures, so that the path over the ground may lie in a given direction and successive photographs may be correctly spaced. It consists essentially of a glass disc, over which swings a ruler pivoted so that lines drawn on the glass along its edge all pass through the pivot. This disc can rotate in a frame and is graduated in degrees around its circumference; it is mounted horizontally in the aeroplane with the line joining its centre to the pivot of the ruler parallel to the direction of motion through the air. When in use, the disc is rotated by hand, until the graduation which points in the direction of motion through the air agrees with the bearing upon which the aeroplane is flying; this results in the disc being always oriented in the same way, no matter what the direction of flight of the aeroplane. A system of sights is attached to the ruler which enables the observer to set it parallel to the direction of motion of the ground relative to the aeroplane, so that, by drawing a line along the edge of this ruler, he can record his direction of motion over the ground.

To use this instrument the pilot flies at a given speed, on a given bearing by compass, and the observer, noting some point on the ground which passes directly beneath him, watches it until it has been left some miles behind and then, using his sights, swings the ruler until the point lies upon a vertical plane passing through it. This is the most convenient way of setting the ruler parallel to the direction of motion over the ground and is called "taking a tail bearing." A line is now drawn on the glass along the edge of the ruler and the whole process is repeated twice more, whilst flying on two other bearings making angles of 120 degrees with the first bearing.

If neither the wind nor the aeroplanes' air-speed have altered during the flights, and if the process has been exactly carried out, the three lines drawn on the glass will intersect in a point. It is not difficult to see that, in effect, three triangles of velocities have been constructed, in which the velocity of the aeroplane through the air is represented by the line joining the centre of the disc to the pivot of the ruler, the velocity over the ground is represented by the lines actually drawn, and the velocity of the wind is represented by the line joining the centre of the disc to the intersection of these three lines. This last line is called the "Wind Vector" and the intersection of the three points is called the "Wind Point."

In practice, owing to slight errors or to changes of wind, the three lines will not meet exactly in a point, but will form a triangle, the size of which gives an indication of the accuracy with which the wind has been determined. If the triangle is not too large its centroid is taken to be the wind point, but if it is too large a fourth flight is made to act as a further check; it is rarely necessary to do this. Once the wind point has been determined it becomes a very simple matter to calculate the bearing and time intervals required for any given photographic flight. This is done by forming yet another triangle of velocities, of which one of the sides is the direction to be made good over the ground, but this triangle cannot be formed directly and a method of trial and error has to be adopted. For this purpose the instrument is equipped with a bar—invisible in the illustration—which lies beneath the glass disc and which can be rotated relatively to it, but which moves with it when not purposely disturbed. This bar is set so that, when the disc is correctly oriented, the bar will be parallel to the direction which has to be made good over the ground. The disc and ruler are then adjusted by trial and error until the ruler both passes through the wind point and is parallel to the bar. This operation sounds more difficult than it is, for it can be performed in a few seconds and requires at most three successive trials.

When this adjustment has been made the course on which the aeroplane has to fly is read off on the circumference of the disc, opposite a fiducial mark, or lubber line, which is placed on the frame, on the line joining the centre of the disc to the pivot of the ruler. The speed of the aeroplane over the ground is then found from the distance between the pivot of the ruler and the wind point; for this distance is proportional to the speed over the ground, on the same scale as the distance from the pivot to the disc centre is proportional to the speed through the air.

When used for navigation the distance from the pivot to the disc centre is adjusted in accordance with the air-speed of the aeroplane, and the speed made good over the ground is read off directly by reference to graduations on the ruler. In aerial surveying, on the other hand, it is seldom necessary to know the ground speed explicitly; it is the time which will be taken to cover some given distance which is required, and to find this we adopt the following procedure.

The distance between the ruler pivot and the disc centre is fixed before leaving the ground, quite irrespective of speed of the aeroplane. Let this distance be d.

Let the distance from the pivot to the wind point be l.

Let T be the time required to cover a distance S on the ground.

Let V_g be the ground speed, V_a be the true air-speed and V_i the air-speed as shown on the speedometer.

Then
$$V_g = l/d \times V_a$$
and
$$T = \frac{S}{V_g} = \frac{d \cdot S}{l \cdot V_a}.$$

If the time interval T is not required with an accuracy greater than 3 or 4 per cent., the true air-speed V_a can be found in terms of the pre-arranged indicated air-speed, by the use of a table of average corrections at various heights*, and the quantity $\dfrac{S \cdot d}{V_a}$ can be calculated before leaving the ground. The observer will then have nothing to do but divide this factor by the length l, as determined on his wind gauge bearing plate.

If the time interval is required to a greater accuracy than can be given by the above procedure, the value of the factor must be determined in the air, after reading the altimeter and the thermometer. To do this we can use the result given in Appendix VI and obtain the relations

$$T = \frac{Sd}{V_a l} = \left[\frac{Sd}{KV_i} \right] \left[\frac{p_0}{b} \right]^{\frac{1}{2}} \left[\frac{\theta_0}{\theta} 10^{-h/A} \right]^{\frac{1}{2}} \times \frac{1}{l}.$$

To deal with this formula in practice we observe that S, d, K and V_i will, in general, be constants which remain unaltered throughout a survey; hence the value of the first bracket can be calculated once for all. The value of the second bracket

* See page 117.

depends upon the barometric reading b at the level of the aerodrome and can, therefore, be ascertained before leaving the ground. The observer therefore leaves the ground with the whole of the first two brackets represented by a single factor. The value of the third bracket can be ascertained in the air, after reading the altimeter and thermometer and referring to a family of curves, which can be drawn out once for all and fixed permanently in the observer's cockpit. These curves might show the value of the bracket as ordinate plotted against temperature as abscissa, each curve referring to a different reading of the altimeter. The observer's form illustrated in chapter VII contains spaces arranged for the convenient reduction of this formula on the ground and in the air. In this form F stands for

the factor $\left[\dfrac{\theta_0}{\theta} 10^{-h/A}\right]^{\frac{1}{2}}$. The first two factors would be ascertained before leaving

the ground and their product entered in the first rectangle on the third line. The observer, when in the air, has to read "h" from his altimeter, "θ" from his thermometer and ascertain "F" from the curves provided; enter "F" in the right-hand rectangle, multiply it by the factor in the left-hand rectangle, enter the result in the bottom rectangle, and divide successively by the two values of l obtained from the wind gauge bearing plate. This gives the two values of "T" which he enters in the appropriate space.

When using the shorter method of calculating "T," the factor in the bottom rectangle can be filled in before leaving the ground.

Printed in the United States
By Bookmasters